Microsoft®
Access®
365 2019 Edition

Nita Rutkosky

Audrey Roggenkamp
Pierce College Puyallup
Puyallup, Washington

Ian Rutkosky
Pierce College Puyallup
Puyallup, Washington

PARADIGM
EDUCATION SOLUTIONS

St. Paul

Vice President, Content and Digital Solutions: Christine Hurney
Director of Content Development: Carley Fruzzetti
Testers: Desiree Carvel; Ann E. Mills, Ivy Tech Community College of Indiana, Evansville, IN
Director of Production: Timothy W. Larson
Production Editor/Project Manager: Jen Weaverling
Cover and Text Design: Valerie King
Senior Design and Production Specialist: Julie Johnston
Copy Editor: Deborah Brandt
Indexer: Terry Casey
Vice President, Director of Digital Products: Chuck Bratton
Digital Projects Manager: Tom Modl
Digital Solutions Manager: Gerry Yumul
Senior Director of Digital Products and Onboarding: Christopher Johnson
Supervisor of Digital Products and Onboarding: Ryan Isdahl
Vice President of Sales: Scott Burns
Vice President of Marketing: Lara Weber McLellan

Care has been taken to verify the accuracy of information presented in this book. However, the authors, editors, and publisher cannot accept responsibility for web, email, newsgroup, or chat room subject matter or content, or for consequences from the application of the information in this book, and make no warranty, expressed or implied, with respect to its content.

Trademarks: Microsoft is a trademark or registered trademark of Microsoft Corporation in the United States and/or other countries. Some of the product names and company names included in this book have been used for identification purposes only and may be trademarks or registered trade names of their respective manufacturers and sellers. The authors, editors, and publisher disclaim any affiliation, association, or connection with, or sponsorship or endorsement by, such owners.

Paradigm Publishing Solutions is independent from Microsoft Corporation and not affiliated with Microsoft in any manner.

Cover Photo Credit: © whitehoune/Shutterstock.com; © Fuatkose/iStock.com.

We have made every effort to trace the ownership of all copyrighted material and to secure permission from copyright holders. In the event of any question arising as to the use of any material, we will be pleased to make the necessary corrections in future printings.

ISBN 978-0-76388-693-6 (print)
ISBN 978-0-76388-687-5 (digital)

© 2020 by Paradigm Publishing, LLC
875 Montreal Way
St. Paul, MN 55102
Email: CustomerService@ParadigmEducation.com
Website: ParadigmEducation.com

Printed in the United States of America

27 26 25 24 23 22 21 20 19   1 2 3 4 5 6 7 8 9 10 11 12

Contents

Access®

Microsoft Access is a *database management system (DBMS)* included with the Microsoft Office suite. Interacting with a DBMS occurs often as one performs daily routines such as withdrawing cash from the ATM, purchasing gas using a credit card, or looking up a telephone number in an online directory. In each of these activities a DBMS is accessed to retrieve information, and data is viewed, updated, and/or printed. Any application that involves storing and maintaining a large amount of data in an organized manner can be set up as an Access database. Examples include customers and invoices, suppliers and purchases, inventory and orders. While working in Access, you will create and maintain databases for the following six companies.

First Choice Travel is a travel center offering a full range of traveling services from booking flights, hotel reservations, and rental cars to offering travel seminars.

The Waterfront Bistro offers fine dining for lunch and dinner and also offers banquet facilities, a wine cellar, and catering services.

Worldwide Enterprises is a national and international distributor of products for a variety of companies and is the exclusive movie distribution agent for Marquee Productions.

Marquee Productions is involved in all aspects of creating movies from script writing and development to filming. The company produces documentaries, biographies, as well as historical and action movies.

Performance Threads maintains an inventory of rental costumes and also researches, designs, and sews special-order and custom-made costumes.

The mission of the Niagara Peninsula College Theatre Arts Division is to offer a curriculum designed to provide students with a thorough exposure to all aspects of the theatre arts.

In Section 1 you will learn how to

Maintain Data in Tables

Access databases are comprised of a series of objects. A table is the first object that is created in a new Access database. Information in the database is organized by topic, and a table stores data for one topic. For example, one table in a customer database might store customer names and addresses while another table stores the customer invoices and yet another table stores the customer payments. Table datasheets are organized like a spreadsheet, with columns and rows. Each column in the table represents one *field*, which is a single unit of information about a person, place, item, or object. Each row in the table represents one *record*, which includes all of the related fields for one person, place, item, or object. Working in tables involves adding or deleting records; editing fields; and sorting, filtering, or formatting datasheets. Access provides the Navigation pane for managing database objects.

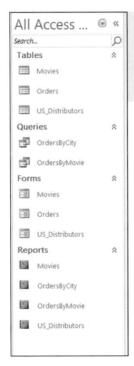

Group objects in the database by various categories and display them in the Navigation pane.

Data in tables display in a datasheet comprised of columns and rows similar to an Excel worksheet. Each column in a table datasheet represents one field. Each row in a table datasheet represents one record.

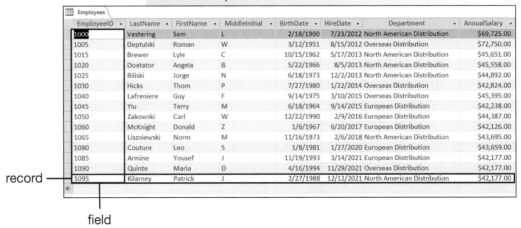

record

field

In Section 2 you will learn how to

Create New Tables and Establish Relationships

New tables can be created starting with a blank datasheet or by creating the table structure by defining fields in a view called *Design view*. Each field in a table has a set of *field properties*, which are characteristics that control how the field interacts with data in objects such as tables, forms, queries, or reports. The ability to create a relationship between two tables allows one to maintain or extract data in multiple tables as if they were one large table.

Create a new table in Design view by assigning each field a field name, a data type, and field properties.

field properties

Designate one field in a table as a primary key—a field that will contain unique data for each record.

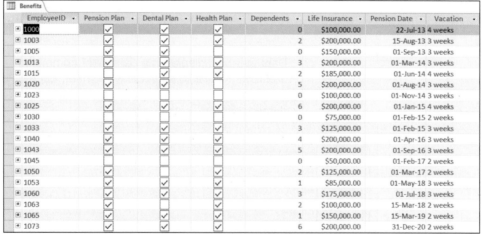

Create relationships between tables by joining one table to another using a common field. These relationships are displayed in the Relationships window using black join lines between table field list boxes.

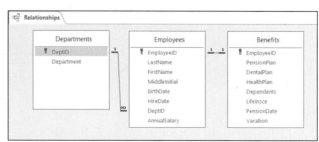

In Section 3 you will learn how to

Create Queries, Forms, and Reports

Queries and forms are objects based on tables and are created to extract, view, and maintain data. Queries can be used to view specific fields from tables that meet a particular criterion. For example, create a query to view customers from a specific state or zip code. Forms provide a more user-friendly interface for entering, editing, deleting, and viewing records in tables. Create a report to generate professionally designed printouts of information from tables or queries.

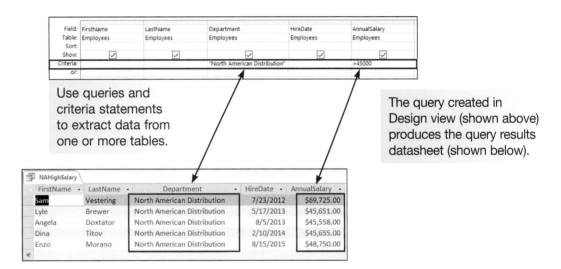

Use queries and criteria statements to extract data from one or more tables.

The query created in Design view (shown above) produces the query results datasheet (shown below).

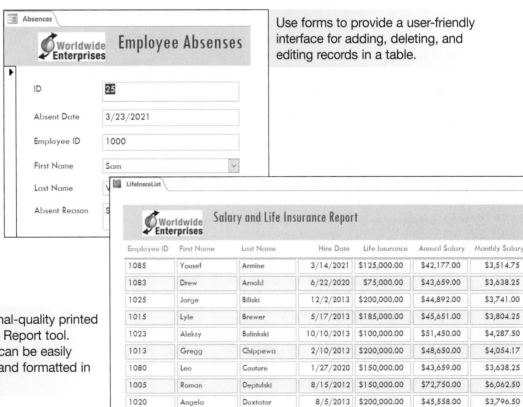

Use forms to provide a user-friendly interface for adding, deleting, and editing records in a table.

Create professional-quality printed reports using the Report tool. Report columns can be easily moved, resized, and formatted in Layout view.

In Section 4 you will learn how to

Summarize Data and Create Calculations

Large amounts of data in tables can be summarized using various tools available within Access. Aggregate functions calculate statistics such as the sum, average, maximum, and minimum values on a numeric field. Crosstab queries use functions on a numeric field to group data by two fields. Forms and reports can have data conditionally formatted, and you can add a calculated field to a form and/or report to calculate a result from a numeric field. Reports can be more meaningful when like data is grouped and sorted. Mailing labels are produced easily with the assistance of the Labels Wizard.

SalaryStatsByDept

SumOfAnnualSalary	AvgOfAnnualSalary	MaxOfAnnualSalary	MinOfAnnualSalary	Department
$314,687.00	$44,955.29	$51,450.00	$42,126.00	European Distribution
$645,975.00	$46,141.07	$69,725.00	$41,875.00	North American Distribution
$203,146.00	$50,786.50	$72,750.00	$42,177.00	Overseas Distribution

Summarize data by calculating statistics such as sum, average, maximum, and minimum. Show statistics for the entire table or group them by a field.

Use a crosstab query to group data by two fields. In this example, annual salary is grouped by department (rows) and by year (columns).

PayrollByDeptByYear

Department	Total Of AnnualSalary	2012	2013	2014	2015	2016	2017	2018	2019	2020	2021
European Distribution	$314,687.00		$100,100.00		$42,238.00	$44,387.00	$42,126.00			$43,659.00	$42,177.00
North American Distribution	$645,975.00	$111,600.00	$136,101.00	$45,655.00	$48,750.00		$43,695.00	$87,390.00	$44,771.00	$43,659.00	$84,354.00
Overseas Distribution	$203,146.00	$72,750.00		$42,824.00	$45,395.00						$42,177.00

Add labels, calculations, and images to forms (as shown at right) and group and sort reports (as shown below).

PDCourses

Worldwide Enterprises

Professional Development Course Reimbursement

EmployeeID	1020
Course Description	Introduction to Accounting
Registration Date	05-Sep-20
End Date	15-Dec-20
Grade	B-
Tuition	$485.00
Accredited School	☑
Transcript Received	☑
Date Reimbursed	23-Jan-21

MINIMUM GRADE OF C- IS REQUIRED

EmployeesByDept

Employee Salaries by Department

Thursday, January 24, 2019
9:31:46 PM

Department	Employee ID	First Name	Last Name	Hire Date	Annual Salary
European Distribution					
	1085	Yousef	Armine	3/14/2021	$42,177.00
	1023	Aleksy	Bulinkski	10/10/2013	$51,450.00
	1013	Gregg	Chippewa	2/10/2013	$48,650.00
	1080	Leo	Couture	1/27/2020	$43,659.00
	1060	Donald	McKnight	6/20/2017	$42,126.00
	1045	Terry	Yiu	9/14/2015	$42,238.00
	1050	Carl	Zakowski	2/9/2016	$44,387.00
North American Distribution					
	1083	Drew	Arnold	6/22/2020	$43,659.00
	1025	Jorge	Biliski	12/2/2013	$44,892.00

Getting Started

Adjusting Monitor Settings, Copying Data Files, and Changing View Options

Skills

- Set monitor resolution
- Modify DPI settings
- Retrieve and copy data files
- Change view options

This textbook and the accompanying eContent were written using a typical personal computer (tower/box, monitor, keyboard and mouse) or laptop. Although you may be able to perform some of the activities in this textbook on a different operating system or tablet, not all the steps will work as written and may jeopardize any work you may be required to turn in to your instructor. No matter what computer you use, you can access the content using the virtual Office experience in Cirrus. If you are unable to access Cirrus or a compatible computer, explore what options you have at your institution such as where and when you can use a computer lab.

One of the evolutions of the Microsoft Office product is that it is offered in a subscription-based plan called Microsoft Office 365. An advantage of having an active Microsoft Office 365 subscription is that the subscription includes and incorporates new features or versions as they are released. This method of providing the Microsoft Office product may impact section activities and assessments. For example, new features and adjustments made to Office 365 may alter how some of the steps discussed in the Marquee Series are completed. The Marquee eContent and online course will contain the most up-to-date material and will be updated as new features become available.

In Activity 1 you will customize your monitor settings so that what you see on the screen should match the images in this textbook. In Activity 2 you will obtain the data files you will be using throughout this textbook from your Cirrus online course. Activity 3 includes instructions on how to change the view settings so that your view of files in a File Explorer window matches the images in this textbook.

Before beginning activities in this textbook, you may want to customize your monitor's settings. Activities in the sections in this textbook assume that the monitor display is set at 1920 x 1080 pixels and the DPI is set at 125%. Adjusting a monitor's display settings is important because the ribbon in the Microsoft Office applications adjusts to the screen resolution setting of your computer monitor. A monitor set at a high resolution will have the ability to show more buttons in the ribbon than a monitor set to a low resolution. Figure GS1 at the bottom of the page shows, the Word ribbon: at the screen resolution featured throughout this textbook and in the virtual office experience (1920 × 1080).

What You Will Do   Adjust the monitor settings for your machine to match the settings used to create the images in the textbook. If using a lab computer, check with your instructor before attempting this activity.

1 Right-click a blank area of the desktop and then click the *Display settings* option at the shortcut menu.

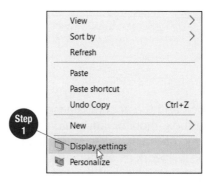

2 At the Settings window with the *Display* option selected, scroll down and look at the current setting displayed in the *Resolution* option box. If your screen is already set to 1920 × 1080, skip ahead to Step 5.

Screen resolution is set in pixels. *Pixel* is the abbreviation of *picture element* and refers to a single dot or point on the display monitor. Changing the screen resolution to a higher number of pixels means that more information can be seen on the screen as items are scaled to a smaller size.

Figure GS1 Word Ribbon Set at 1920 x 1080 Screen Resolution

3 Click the *Resolution* option box and then click the *1920 × 1080* option. ***Note:*** ***Depending on the privileges you are given on a school machine, you may not be able to complete Steps 3–7. If necessary, check with your instructor for alternative instructions.***

> If the machine you are using has more than one monitor, make sure the proper monitor is selected. (The active monitor displays as a blue rectangle in the Display pane of the Settings app.)

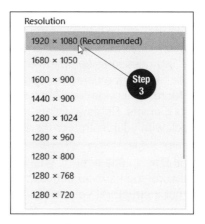

4 Click the Keep changes button at the message box asking if you want to keep the display settings.

> Some monitor settings will render the computer unusable because objects on the desktop or in a window will become inaccessible and hidden. In this case, Windows will automatically revert the settings to the previous configuration after 30 seconds.

5 At the Settings window with the *Display* option active, look at the percentage in which the size of text, apps, and other items currently display (also known as the DPI setting). For example, items on your screen may display at 100%. If the percentage is 125%, skip to Step 12.

> The computers used to create the images in this textbook uses the 125% DPI setting, which slightly increases the size of text, applications, buttons, and options.

6 Click the option box below the text *Change the size of text, apps, and other items*, and then click the *125%* option in the drop-down list.

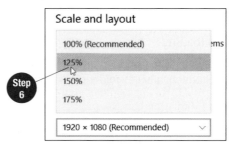

7 The message *Some apps won't respond to scaling changes until you sign out.* appears in the *Scale and Layout* section of the dialog box. You may need to sign out of Windows and restart your computer for the scaling options to apply.

Activity 2 Retrieving and Copying Data Files

While working through the activities in this book, you will often be using data files as start-ing points. These files need to be obtained from your Cirrus online course or other locations such as your school's network drive. All the files required to complete the coursework are provided through Cirrus. You have the ability to access all of the data files from the Course Resources. You may download all of the files at once or only download the files you need for a specific section (described in the activity below). Cirrus online course activities also provide the individual files needed to complete each activity. Make sure you have internet access before trying to retrieve the data files from Cirrus. Ask your instructor if alternate locations are available for retrieving the files, such as a network drive or online resource such as D2L, BlackBoard, or Canvas. Retrieving data files from an alternate location will require different steps, so check with your instructor for additional steps or tasks to complete.

What You Will Do   In this activity, you will download data files from your Cirrus online course. Make sure you have an active internet connection before starting this activity. Check with your instructor if you do not have access to your Cirrus online course.

1 Insert your USB flash drive into an available USB port.

2 Navigate to the Course Resources section of your Cirrus online course. *Note: The steps in this activity assume you are using the Chrome browser. If you are using a different browser, the following steps may vary.*

3 Click the Student Data Files link in the Course Resources section.

A zip file containing the student data files will automatically begin downloading from the Cirrus website.

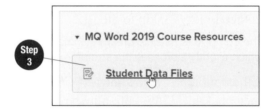

4 Click the button in the lower left corner of the screen once the files have finished downloading.

Clicking the button in the lower left corner of the screen will open File Explorer and the StudentDataFiles folder displays in the Content pane.

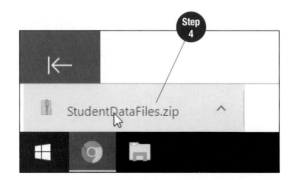

5 Right-click the *StudentDataFiles* folder in the Content pane.

6 Click the *Copy* option at the shortcut menu.

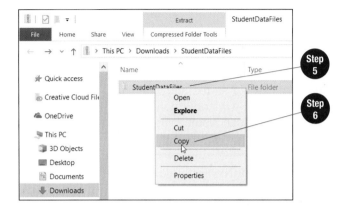

7 Click your USB flash drive that displays in the Navigation pane at the left of the File Explorer window.

8 Click the Home tab and then click the Paste button in the Clipboard group.

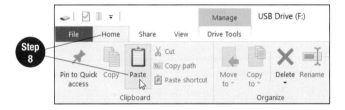

9 Close the File Explorer window by clicking the Close button in the upper right corner of the window.

You can change the view of the File Explorer window to show the contents of your current location (drive or folder) in various formats, including icons, tiles, or a list, among others. With the Content pane in Details view, you can click the column headings to change how the contents are sorted and whether they are sorted in ascending or descending order. You can customize a window's environment by using buttons and options on the File Explorer View tab. You can also change how panes are displayed, how content is arranged in the Content pane, how content is sorted, and which features are hidden.

What You Will Do   Before getting started with the textbook material, you need to adjust the view settings so that items in the File Explorer window appear the same as the images in the textbook.

1   Click the File Explorer button on the taskbar.

> By default, a File Explorer window opens at the Quick access location, which contains frequently-used folders such as Desktop, Documents, Downloads, Pictures and so on. It also displays recently used files at the bottom of the Content pane.

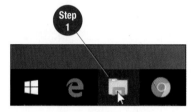

2   Click the drive letter representing your storage medium in the Navigation pane.

3   Double-click the *ExcelS2* folder in the Content pane.

4   Click the View tab below the Title bar.

5   Click the *Large icons* option in the Layout group.

> After you click an option on the View tab, the View tab collapses to provide more space in the File Explorer window.

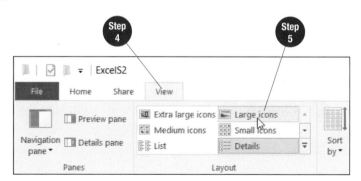

6 Click the View tab.

7 Click the *Details* option in the Layout group.

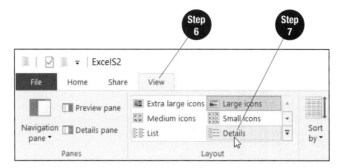

8 With files now displayed in Details view, click the *Name* column heading to sort the list in descending order by name.

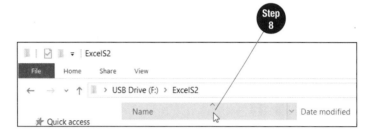

9 Click the *Name* column heading again to restore the list to ascending order by name.

10 Close the File Explorer window by clicking the Close button in the upper right corner of the window.

In Addition

Changing the Default View for All Folders

You can set a view to display by default for all folders of a similar type (such as all disk drive folders or all documents folders). To do this, change the current view to the desired view for the type of folder that you want to set. Next, click the Options button on the View tab and then click the View tab at the Folder Options dialog box. Click the Apply to Folders button in the Folder views section and then click OK. Click Yes at the Folder Views message asking if you want all folders of this type to match this folder's view settings.

Turning on File Extensions

By default, file extensions do not display. Turning on file extensions is helpful in identifying the type of file. Turn on file extensions at the File Explorer window by clicking the View tab and then clicking the *File name extensions* check box to insert a check mark.

Maintaining Data in Access Tables

 Data Files

Before beginning section work, copy the AccessS1 folder to your storage medium and then make AccessS1 the active folder.

Skills

- Open and close Access
- Open and close an existing database
- Open and close tables, queries, forms, and reports
- Explore and navigate in tables
- Adjust field column widths
- Find and replace data
- Add and delete records in a table and form

- Sort and filter records
- Preview and print a table
- Change margins
- Change the page orientation
- Use the Help and Tell Me features
- Change the font size for records in a table
- Hide columns in a table

Projects Overview

 **Worldwide Enterprises**

Add, delete, find, sort, and filter records; change page orientation and margins; format and hide columns in a table in two databases.

 **NIAGARA PENINSULA COLLEGE**

Find student records and enter grades into the Grades database.

 **The Waterfront BISTRO**

Maintain the Inventory database by adding and deleting records. Add, delete, and modify records; sort, filter, and set print options for a Catering Event database.

 **Performance Threads**

Delete records and sort and print a report from the Costume Inventory database. Create field names and table names for a new Custom Costume database.

Note: On some computer systems, a file downloaded from the internet, or copied from a storage medium retains the source file's read-only attribute. If this occurs, you will not be able to make changes to the database file. Complete the following steps to check and then remove a file's read-only status:

1. Open a File Explorer window.
2. Navigate to the drive representing your storage medium.
3. Navigate to the AccessS1 folder.
4. Right-click a database file and then click *Properties* at the shortcut menu.
5. At the Properties dialog box with the General tab selected, look at the *Read-only* check box in the *Attributes* section. If it contains a check mark, click the check box to remove it; if the check box is empty, the read-only attribute is not turned on (no action is required).
6. Click OK.

If the read-only attribute is not active for one file in the copied folder, then you can assume all files copied without an active read-only attribute. If you cleared the *Read-only* check box for one file, you will need to clear the attribute for all files after downloading from Cirrus. You can select multiple files and remove the read-only attribute in one operation.

 The online course includes additional training and assessment resources.

Organizations use databases to keep track of customers, suppliers, employees, inventory, sales, orders, purchases, and much more. A database can be defined as a collection of data that has been organized so that the data can be easily stored, sorted, extracted, and reported. A key concept for understanding databases is that the data has to be organized. Data is organized first into a series of tables within the database, where one table contains all of the data that describe a person, place, object, event, or other subject. Within a database, a series of objects exist for entering, managing, and viewing data. The first objects created are *tables*. Once a table exists, other objects can be created that use the table structure as a means to enter and view the data. Some other database objects include queries, forms, and reports.

Worldwide Enterprises

What You Will Do   You will open a database used by Worldwide Enterprises to keep track of distributors, orders, and movies and open and close the objects in the database to gain an understanding of important Access concepts and terminology.

Note: Make sure you have downloaded the files from Cirrus to your storage medium. Please refer to the note on page 1 regarding checking and removing read-only attributes before proceeding with the activities in this section.

Tutorial
Opening an Existing Database

Tutorial
Opening and Closing an Object

Tutorial
Closing a Database and Closing Access

1 At the Windows desktop, click the Start button and then click the Access tile.

Depending on your system configuration, these steps may vary.

2 At the Access opening screen, click the <u>Open Other Files</u> hyperlink at the bottom of the Recent list.

3 At the Open backstage area, click the *Browse* option.

4 At the Open dialog box, click *Removable Disk (F:)* in the Navigation pane.

If you are opening and saving files in a location other than a USB flash drive, please check with your instructor. You can bypass the Open backstage area and go directly to the Open dialog box by pressing Ctrl + F12.

5 Double-click the *AccessS1* folder in the Content pane of the Open dialog box.

6 Double-click *1-WEDistributors* in the Content pane.

A security warning message bar displays if Access determines the file you are opening did not originate from a trusted location on your computer. This often occurs when you copy a file from another medium (such as a flash drive or the Web).

7 If a security warning message bar displays, click the Enable Content button.

Identify the various features of the Access screen by comparing your screen with the one shown in Figure 1.1.

Step 6

Figure 1.1 The Access Screen

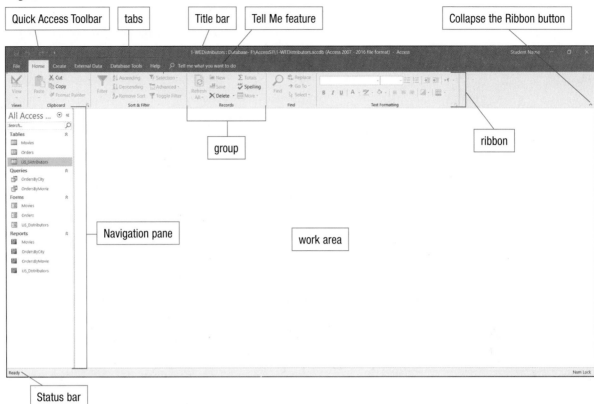

Quick Access Toolbar | tabs | Title bar | Tell Me feature | Collapse the Ribbon button

group | ribbon

Navigation pane | work area

Status bar

8 Double-click the *Orders* table in the Tables group in the Navigation pane.

The Orders table opens in Datasheet view in the work area with a tab in the upper left corner identifying the table name. Each column in the table contains only one unit of information and each row represents one record. A table is an object in a database and other objects are created using data from a table or multiple tables. Refer to Table 1.1 for a description of database objects.

Step 8

Table 1.1 Database Objects

Object	Description
table	organizes data in fields (columns) and records (rows); a database must contain at least one table. The table is the base upon which other objects are created
query	displays data from one or more tables that meets a conditional statement (for example, records in which the city is Toronto); performs calculations
form	allows fields and records to be presented in a different layout than the datasheet; used to facilitate data entry and maintenance
report	prints data from tables or queries
macro	automates repetitive tasks
module	advances automation through programming using Visual Basic for Applications

9 Double-click the *OrdersByCity* query in the Queries group in the Navigation pane.

> The OrdersByCity query opens in a new tab in the work area. A query resembles a table in that the information displays in a column-and-row format. The purpose of a query is to display data from one or more related tables that meets a conditional statement.

10 Double-click the *Movies* form in the Forms group in the Navigation pane. (Make sure you double-click *Movies* in the Forms group and not one of the other groups.)

> The Movies form opens in a new tab in the work area. A form presents fields and records in a layout different from a table and is used to facilitate data entry and maintenance.

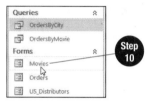

11 Right-click the *US_Distributors* report in the Reports group in the Navigation pane. (Make sure you right-click *US_Distributors* in the Reports group and not one of the other groups.)

> Right-clicking the report displays the shortcut menu. The shortcut menu provides another method for opening an object.

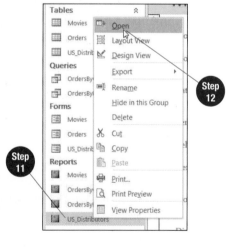

12 Click *Open* at the shortcut menu.

> The US_Distributors report opens in a new tab in the work area. Reports are used to display or print data from one or more tables in a specific layout. In a report, data can be arranged, formatted, grouped, and sorted. Descriptive labels, a logo, or other items can be included.

13 Click the Orders tab in the work area.

> Clicking a tab in the work area moves that object to the foreground.

14 Close the Orders table by clicking the Close button in the upper right corner of the work area.

In Brief

Open Database
1. Click Open Other Files hyperlink.
2. Click *Browse* option.
3. Click *Removable Disk (F:)*.
4. Navigate to folder.
5. Double-click database file.
OR
1. Click File tab and then click *Open*.
2. Click *Browse* option.
3. Click *Removable Disk (F:)*.
4. Navigate to folder.
5. Double-click database file.

Open Object
Double-click object in Navigation pane.
OR
1. Right-click object.
2. Click *Open*.

Close Object
Click Close button.
OR
1. Right-click object tab.
2. Click *Close*.

15 Close the US_Distributors report by right-clicking the tab and then clicking *Close* at the shortcut menu.

The shortcut menu provides another method for closing an object.

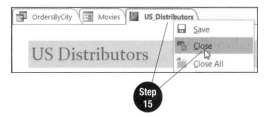

16 Close both of the two open objects by right-clicking the Movies tab and then clicking *Close All* at the shortcut menu.

17 Close the **1-WEDistributors** database by clicking the File tab and then clicking the *Close* option.

Clicking the File tab displays the backstage area with options for working with and managing databases.

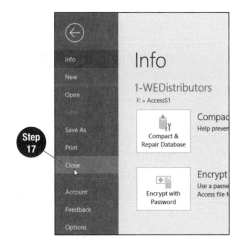

18 Close Access by clicking the Close button in the upper right corner of the screen.

In Addition

Learning More about the Access User Interface

While several files of the same type can be opened in the other Microsoft Office applications, only one Access database can be open at a time. If a database is open and then another one is opened, the first database closes. In other Microsoft Office applications, revisions, such as editing text or values, must be saved. In a database, changes to the data are saved automatically when the insertion point is moved to the next record. For this reason, the Undo command is often unavailable. If a prompt to save changes displays when closing an object in Access, the prompt is referring to changes made to the layout and/or formatting of the object.

In a database, tables are the first objects created, since all other objects rely on tables as the source for their data. Each table in a database contains data for one subject such as customers, orders, suppliers, and so on. Within a table, the data is broken down into units of information about the subject, called *fields*. All of the data about one subject in the table is called a *record*. For example, in a table containing data on suppliers, one record contains the data for one supplier. Navigate in an open table using the mouse, keys on the keyboard, and/or the Record Navigation bar at the bottom of the work area.

Worldwide Enterprises

What You Will Do   You will open the database used by Worldwide Enterprises and examine the tables to determine how information is organized. You will also navigate to various fields and records in the tables.

Tutorial
Navigating in Objects

1 Open Access and then open the **1-WEDistributors** database from your AccessS1 folder and enable the content, if necessary.

2 Double-click the *US_Distributors* table in the Navigation pane.

> Each column in the table in Datasheet view contains only one unit of information. Each column represents one field. Identify the fields and field names at the top of each column by comparing your screen with the one shown in Figure 1.2. The field names *DistID*, *CompanyName*, *StreetAdd1*, and so on display in the header row at the top of the table.

3 Double-click the *Movies* table in the Navigation pane.

> The Movies table opens in the work area on top of the US_Distributors table. Examine the fields and records in the table to determine how the information is organized.

4 Double-click the *Orders* table in the Navigation pane.

> The Orders table opens in the work area on top of the other two tables. Database designers often create a visual representation of the database's structure in a diagram similar to the one in Figure 1.3. Each table is represented in a box with the table name at the top of the box. Within each box, the fields that will be stored in the table are listed. The field with the asterisk is called the *primary key field*, which holds the data that uniquely identifies each record in the table (usually an identification number).

Figure 1.2 US_Distributors Table in Datasheet View

	DistID	CompanyNar	StreetAdd1	StreetAdd2	City	State	ZIPCode	Telephone	Fax	EmailAdd	Click to Add
	1	All Nite Cinema	2188 Third Stree		Louisville	KY	40201	502-555-4238	502-555-4240	allnite@ppi-edu	
	2	Century Cinema	3687 Avenue K		Arlington	TX	76013	817-555-2116	817-555-2119	centurycinemas	
	3	Countryside Cin	22 Hillside Stree		Bennington	VT	05201	802-555-1469	802-555-1470	countryside@pp	
	4	Danforth Cinem	PO Box 22	18 Pickens Stree	Columbia	SC	29201	803-555-3487	803-555-3421	danforth@ppi-e	
	5	Eastown Movie	PO Box 722	1 Concourse Av	Cambridge	MA	02142	413-555-0981	413-555-0226	eastown@ppi-e	
	6	Hillman Cinema	55 Kemble Aven		Baking Ridge	NJ	07920	201-555-1147	201-555-1143	hillman@ppi-ed	
	7	LaVista Cinema	111 Vista Road		Phoenix	AZ	86355-6014	602-555-6231	602-555-6233	lavista@ppi-edu	
	8	Liberty Cinemas	PO Box 998	12011 Ruston W	Atlanta	GA	73125	404-555-8113	404-555-2349	libertycinemas@	
	9	Mainstream Mc	PO Box 33	333 Evergreen E	Seattle	WA	98220-2791	206-555-3269	206-555-3270	mainstream@pp	
	10	Marquee Movie	1011 South Alar		Los Angeles	CA	90045	612-555-2398	612-555-2377	marqueemovies	
	11	Midtown Movie	1033 Commerci		Emporia	KS	66801	316-555-7013	316-555-7022	midtown@ppi-e	
	12	Mooretown Mc	PO Box 11	331 Metro Place	Dublin	OH	43107	614-555-8134	614-555-8330	mooretown@pp	
	13	O'Shea Movies	59 Erie		Oak Park	IL	60302	312-555-7719	312-555-7381	oshea@ppi-edu	
	14	Redwood Cinem	PO Box 112F	336 Ninth Stree	Portland	OR	97466-3359	503-555-8641	503-555-8633	redwoodcinem	
	15	Sunfest Cinema	341 South Fourt		Tampa	FL	33562	813-555-3185	813-555-3177	sunfest@ppi-ed	
	16	Victory Cinema	12119 South 23		San Diego	CA	97432-1567	619-555-8746	619-555-8748	victory@ppi-ed	
	17	Waterfront Cine	PO Box 3255		Buffalo	NY	14288	716-555-3845	716-555-3947	waterfrontcine	
	18	Wellington 10	1203 Tenth Sou		Philadelphia	PA	19178	215-555-9045	215-555-9048	wellington10@	
	19	Westview Movi	1112 Broadway		Newark	NJ	10119	212-555-4875	212-555-4877	westview@ppi-	
*	(New)										

Each object opens in a tab in the work area.

field name

Each row is one record in the table.

Each column represents a field in the table.

Record Navigation bar

Record: 1 of 19   No Filter   Search

Figure 1.3 Database Diagram for 1-WEDistributors Database

5 Change how objects display in the Navigation pane by clicking the menu bar at the top of the Navigation pane and then clicking *Tables and Related Views* at the drop-down list.

> Originally, the Navigation pane displayed the objects grouped by type (table, query, form, and report). Clicking *Tables and Related Views* displays each table in the Navigation pane followed by the objects related to the table.

6 Show only the Orders table and related objects by clicking the Navigation pane menu bar and then clicking *Orders* in the *Filter By Group* section of the drop-down list.

7 Redisplay by object type by clicking the button on the menu bar containing a down-pointing triangle in a circle and then clicking *Object Type* at the drop-down list.

> Clicking *Object Type* displays *All Access Objects* in the *Filter By Group* section of the drop-down list. Display the Navigation pane drop-down list by clicking the menu bar or by clicking the button on the menu bar. The name of the button changes depending on what is selected.

8 To view more of the open objects in the work area, click the Shutter Bar Open/Close Button in the upper right corner of the Navigation pane.

> The Navigation pane collapses to a bar at the left of the screen with the text *Navigation Pane* displayed vertically.

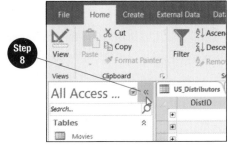

9 Redisplay the Navigation pane by clicking the Shutter Bar Open/Close Button.

> You can also collapse or redisplay the Navigation pane by pressing the F11 function key.

10 Close the Orders table by right-clicking the Orders tab and then clicking *Close* at the shortcut menu.

11 Close the Movies table by right-clicking the Movies tab and then clicking *Close* at the shortcut menu.

The US_Distributors table should be the only open table.

12 Press the Tab key to make the next field in the current record active.

Navigate in a table using keys on the keyboard, the mouse, or the Record Navigation bar. Pressing the Tab key makes the next field in the current record active. Refer to Table 1.2 for information on keyboard commands for navigating in a table.

13 Press Shift + Tab to make the previous field in the current record active.

14 Press Ctrl + End to make the last field in the last record active.

15 Press Ctrl + Home to make the first field in the first record active.

16 Move the insertion point to a field in the *CompanyName* field column by clicking in the company name *Century Cinemas* in record 2.

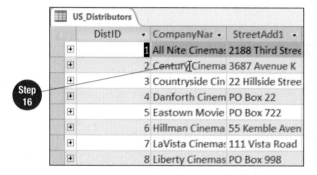

17 Make record 3 active by clicking the Next record button on the Record Navigation bar.

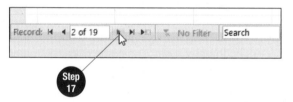

18 Make the last record active by clicking the Last record button on the Record Navigation bar.

Table 1.2 Navigating in an Access Table Using the Keyboard

Press	To move to
Home	first field in the current record
End	last field in the current record
Tab	next field in the current record
Shift + Tab	previous field in the current record
Ctrl + Home	first field in the first record
Ctrl + End	last field in the last record

In Brief

Change Navigation Pane View
1. Click All Access Objects button at top of Navigation pane.
2. Click view.

19 Make the first record active by clicking the First record button on the Record Navigation bar.

Step 19

20 Navigate to record 14 by triple-clicking the text *1 of 19* that displays in the *Current Record* text box on the Record Navigation bar, typing 14, and then pressing the Enter key.

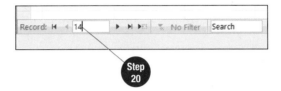

Step 20

21 Close the US_Distributors table by clicking the Close button in the upper right corner of the work area.

In Addition

Planning and Designing a Table

The table below describes the steps involved in planning and designing a new database. The design process may seem time-consuming; however, the time expended to produce a well-designed database saves time later. A database that is poorly designed will likely have logical or structural errors that require redefining of data or objects after live data has been entered.

Step	Description
1. Determine the purpose of the database.	Describe who will use the database and for what purpose. This helps to focus the efforts for the remaining steps on the mission the database is to fulfill.
2. Determine all of the data that will need to be tracked in the database.	Gather all of the data elements that you will need to store in the database. You can find this information by looking at records of invoices, inventory lists, purchase orders, and so on. You can also ask individuals what information they want to get out of the database to help you determine the required data.
3. Group the data elements into tables.	Divide the information into subjects (also referred to as *entities*) so that one table will be about one subject only.
4. Decide the fields and field names for each table.	Break down each data element into its smallest unit. For example, a person's name could be broken down into first name, middle name, and last name.
5. Make sure each table includes a field that will be used to uniquely identify each record in the table.	Access can create an *ID* field for you if you do not have an existing unique identification number such as a product number, student number, social security number, or similar field.
6. Decide which tables need to be linked by a relationship and include in the table the common field upon which to join the tables if necessary.	Identifying relationships at this stage helps you determine if you need to add a field to a related table to allow you to join the table. You will learn more about relationships in Section 2.

Access creates columns in a table with a default width of 13.1111 characters, which is the approximate number of characters that will display in a column. Column width can be increased or decreased to accommodate the data with options at the Column Width dialog box, by double-clicking the column boundary line to adjust the width to the longest entry (referred to as best fit), or by dragging a column boundary line to the desired position. The width of multiple adjacent columns can be best fit by selecting the columns and then clicking one of the column boundary line.

Worldwide Enterprises

What You Will Do   You will reopen the US_Distributors table and adjust column widths to accommodate the longest entries in each column.

Tutorial

Adjusting Field
Column Width

1   With the **1-WEDistributors** database open, double-click the US_Distributors table in the Navigation pane.

2   With the first field selected in the *DistID* field column, click the More button 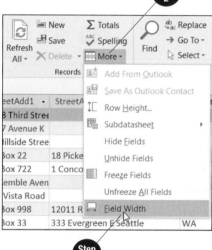 in the Records group on the Home tab.

3   Click *Field Width* at the drop-down list.

4   At the Column Width dialog box with the current width selected in the *Column Width* measurement box, type 8 and then click OK.

> If the data entered in a field in a column does not fill the entire field, consider decreasing the column width to save space in the work area and when the table is printed.

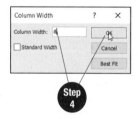

Step 4

5   Click in the first field in the *CompanyName* field column (the field containing the text *All Nite Cinemas*).

> Clicking in a field in the *CompanyName* field column makes the column active and *CompanyName* in the header row displays with an orange background.

6   Click the More button in the Records group and then click *Field Width* at the drop-down list.

7   At the Column Width dialog box with the current width selected in the *Column Width* measurement box, type 20 and then click OK.

8   Click in the first field in the *StreetAdd1* field column (the field containing the text *2188 Third Street*).

9   Click the More button and then click *Field Width* at the drop-down list.

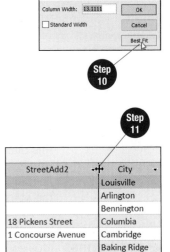

In Brief

Adjust Column Width
1. Position insertion point in column.
2. Click More button.
3. Click *Field Width*.
4. Type column width number or click Best Fit button.
OR
Drag or double-click right column boundary line in header row.

10 At the Column Width dialog box, click the Best Fit button.

Clicking the Best Fit button causes the width of the *StreetAdd1* column to adjust to accommodate the longest entry.

11 Use the mouse to adjust the *StreetAdd2* field column. To do this, position the mouse pointer on the column boundary line between the *StreetAdd2* and *City* field column headings until the pointer displays as a left-and-right pointing arrow with a vertical line between, click and hold down the left mouse button, drag to the right approximately one-half inch, and then release the mouse button. Make sure all of the entries in the column are visible. If not, continue to increase the width of the column.

12 Select the remaining columns by positioning the mouse pointer in the *City* field column heading, clicking and holding down the left mouse button, dragging to the right to the *EmailAdd* field column heading, and then releasing the mouse button.

13 With the columns selected, double-click the column boundary line between the *City* and *State* field column headings.

Double-clicking the column boundary line best fits all of the selected columns. Double-check that the *EmailAdd* field column has fully expanded to display all entries. If not, double-click the column boundary line between the *Fax* and *EmailAdd* field column headings. You may need to scroll to the right to display the entire field.

14 Click the First record button on the Record Navigation bar.

15 Click the Save button on the Quick Access Toolbar.

Clicking the Save button saves the changes made to the column widths.

16 Close the US_Distributors table.

Check Your Work Compare your work to the model answer available in the online course.

In Addition

Using the Record Navigation Bar Search Text Box

Use the Record Navigation bar search text box to search for a specific entry in a field in a table. To use the search text box, click in the text box and then begin typing the search data. As text is typed, text that matches what is typed is selected in the table, which is referred to as *word-wheeling*. For example, to search for Mooretown Movies in the US_Distributors table, start by clicking at the beginning of the first field below the *CompanyName* field column heading, clicking in the search text box, and then typing *m*. This causes Access to select the letter *m* in *Cinemas* in the first entry in the *CompanyName* field column. Type the letter *o* and Access selects *Mo* in the *Eastown Movie House* entry. Type the letter *o* again and Access selects *Moo* in *Mooretown Movies*.

Use the Find and Replace dialog box with the Find tab selected to search for specific data in a table. Display this dialog box with the Find button on the Home tab. Using the Find and Replace dialog box saves time when searching for data when the table contains many records that are not all visible on the screen. Use the Find and Replace dialog box with the Replace tab selected to find specific data in a table and replace it with other data. Display this dialog box with the Replace button on the Home tab. Use options at the Find and Replace dialog box with either tab selected to specify where to search, such as the current field or the entire table, and what to search, such as any part of a field, the whole field, or the start of a field.

Worldwide Enterprises

What You Will Do   You have received a note from Sam Vestering that the city for Westview Movies was entered incorrectly and should be *New York* instead of *Newark*. He also indicated that the post office reference in the addresses should appear as *P.O.* and not *PO*. You will use the Find and Replace dialog box to locate and make the changes.

Tutorial
Finding Data

Tutorial
Finding and Replacing Data

1 With **1-WEDistributors** open, open the US_Distributors table and then click the first field in the *City* field column (contains the text *Louisville*).

2 Click the Find button [🔍] in the Find group on the Home tab.

Clicking the Find button displays the Find and Replace dialog box with the Find tab selected.

3 Type Newark in the *Find What* text box.

4 Click the Find Next button.

The city *Newark* is selected in record 19.

5 Click the Cancel button to close the Find and Replace dialog box.

6 With *Newark* selected, type New York.

As you type *New York* a pencil icon displays in the record selector bar at the left of the table, indicating that the current record is being edited and the change has not been saved. When you complete Step 7, the change is saved and the pencil icon disappears.

7 Click in the first field in the *DistID* field column (contains the number *1*).

8 Click the Replace button 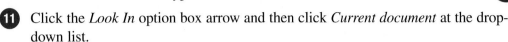 in the Find group on the Home tab.

9 At the Find and Replace dialog box with the Replace tab selected, type PO in the *Find What* text box.

10 Press the Tab key to move the insertion point to the *Replace With* text box and then type P.O.

11 Click the *Look In* option box arrow and then click *Current document* at the drop-down list.

> Changing the *Look In* option to *Current document* tells Access to search all fields in the table.

12 Click the *Match* option box arrow and then click *Any Part of Field* at the drop-down list.

> Changing the *Match* option to *Any Part of Field* tells Access to search for data that matches any part of a field—not only the entire field.

13 Click the *Match Case* check box to insert a check mark.

> Inserting a check mark in the *Match Case* check box tells Access to search only for the uppercase letters *PO*.

14 Click the Replace All button.

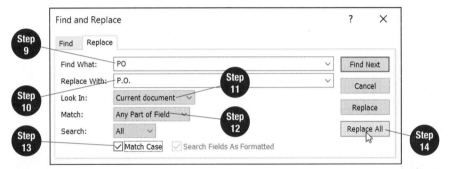

15 At the message indicating that you cannot undo the action and asking if you want to continue, click Yes.

> Clicking the Yes button closes the message and replaces all occurrences of *PO* with *P.O.*

16 Click the Cancel button to close the Find and Replace dialog box.

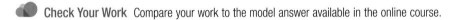 **Check Your Work** Compare your work to the model answer available in the online course.

In Addition

Exploring the Find and Replace Dialog Box Options

By default, the *Look In* option at the Find and Replace dialog box is set to *Current field*. At this setting, Access will search only in the field where the insertion point is positioned. Change this option to *Current document* and Access will search all fields in the table. The *Match* option has a default setting of *Whole Field*, which means Access will search for data that matches the entire entry in a field. To search for data that matches only partial field entries, change the *Match* option to *Any Part of Field* or *Start of Field* to search entries in

fields that begin with the data entered in the *Find What* text box. The *Search* option has a default setting of *All*, which means Access will search all of the data in the specific field or table. This can be changed to *Up* to search from the insertion point to the beginning of the table or *Down* to search from the insertion point to the end of the table. To find data that contains specific uppercase or lowercase letters, insert a check mark in the *Match Case* check box.

Tables are the first objects created in a database and all other objects rely on a table as the source for their data. Managing the tables in a database is important for keeping the database up to date and may include adding and deleting records. Add a new record to a table in Datasheet view by clicking the New (blank) record button on the Record Navigation bar, or clicking the New button in the Records group on the Home tab. Delete a record in a table by clicking in any field in the record, clicking the Delete button arrow in the Records group on the Home tab, and then clicking *Delete Record* at the drop-down list. Another method for deleting a record is to select the record by clicking the record selector bar at the left of the record and then clicking the Delete button on the Home tab.

Worldwide Enterprises

What You Will Do   Worldwide Enterprises has signed two new distributors in the United States and one company has signed with another movie distributing company. You will add the new information in two records in the US_Distributors table and delete the old company.

Tutorial

Adding and Deleting
Records in a Table

1 With 1-WEDistributors open and the US_Distributors table open, click the New (blank) record button ▶ in the Record Navigation bar.

> The insertion point moves to the first field in the blank row at the bottom of the table in the *DistID* field column. The *Current Record* text box in the Record Navigation bar indicates you are editing record 20 of 20 records.

2 Press the Tab key to move past the *DistID* field (currently displays *(New)*) since Access automatically assigns the next sequential number to this field.

> You will learn more about the AutoNumber data type field in Section 2. The number will not appear in the field until you type an entry in the next field.

3 Type Dockside Movies in the *CompanyName* field and then press the Tab key.

4 Type P.O. Box 224 and then press the Tab key.

5 Type 155 Central Avenue and then press the Tab key.

6 Type Baltimore and then press the Tab key.

7 Type MD and then press the Tab key.

8 Type 21203 and then press the Tab key.

9 Type 301-555-7732 and then press the Tab key.

10 Type 301-555-9836 and then press the Tab key.

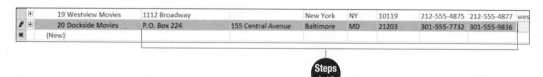

In Brief

Add Records to Datasheet
1. Open table.
2. Click New (blank) record button in Navigation bar or click New button in Records group.
3. Type data in fields.

11 Type dockside@ppi-edu.net and then press the Tab key.

> The insertion point moves to a new row when you press the Tab key or the Enter key after the last field in a new record to allow you to continue typing the next new record in the table. The record just entered is saved automatically.

12 Type the following information in the appropriate fields in the next row:

> Renaissance Cinemas
> 3599 Woodward Avenue
> Detroit, MI 48211
> 313-555-1693
> 313-555-1699
> rencinemas@ppi-edu.net

14 Redwood Cinemas	P.O. Box 112F	336 Ninth Street	Portland	OR	97466-3359	503-555-8641	503-555-8633	redwoodcinemas@ppi-edu.net
15 Sunfest Cinemas	341 South Fourth Avenue		Tampa	FL	33562	813-555-3185	813-555-3177	sunfest@ppi-edu.net
16 Victory Cinemas	12119 South 23rd		San Diego	CA	97432-1567	619-555-8746	619-555-8748	victory@ppi-edu.net
17 Waterfront Cinemas	P.O. Box 3255		Buffalo	NY	14288	716-555-3845	716-555-3947	waterfrontcinemas@ppi-edu.net
18 Wellington 10	1203 Tenth Southwest		Philadelphia	PA	19178	215-555-9045	215-555-9048	wellington10@ppi-edu.net
19 Westview Movies	1112 Broadway		New York	NY	10119	212-555-4875	212-555-4877	westview@ppi-edu.net
20 Dockside Movies	P.O. Box 224	155 Central Avenue	Baltimore	MD	21203	301-555-7732	301-555-9836	dockside@ppi-edu.net
21 Renaissance Cinemas	3599 Woodward Avenue		Detroit	MI	48211	313-555-1693	313-555-1699	rencinemas@ppi-edu.net
(New)								

Step 12

13 Position the mouse pointer in the record selector bar (empty column to the left of the *DistID* field) for record 3 (*Countryside Cinemas*) until the pointer changes to a right-pointing black arrow and then click the left mouse button.

> This action selects the entire record.

14 Click the Delete button ☒ in the Records group on the Home tab.

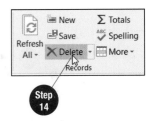

15 Access displays a message box indicating you are about to delete 1 record and that the undo operation is not available after this action. Click the Yes button to confirm the deletion.

Step 14

> Access does not renumber the remaining records in the *DistID* field once record 3 has been deleted from the table. The *DistID* field is defined as an AutoNumber data type field. For this field data type, once a number has been used, Access does not make the number available again for another record even after the record is deleted.

16 Close the US_Distributors table.

 Check Your Work Compare your work to the model answer available in the online course.

In Addition

Understanding the Primary Key Field

In each table, one field is designated as the primary key field. A primary key field is the field by which the table is sorted whenever the table is opened. The primary key field must contain unique data for each record. When a new record is added to a table, Access checks to ensure that no record exists with the same data in the primary key field. If a record does exist with the same data in the primary key field, Access displays an error message indicating there are duplicate values and will not allow the record to be saved. The primary key field cannot be left blank when a new record is added, since it is the field that is used to sort and check for duplicates. Access includes a feature where a field named *ID* defined as the primary key field is included automatically in a new table that is created in a blank datasheet. The *ID* field uses the AutoNumber data type, which assigns the first record a field value of *1* and each new record the next sequential number.

Forms are used to enter, edit, view, and print data. Adding records in a form is easier than using a table since all of the fields in the table are presented in a different layout, which usually allows all fields to be visible at the same time. Other records in the table do not distract the user since only one record displays at a time. Add records to a form using the New (blank) record button on the Record Navigation bar or with the New button in the Records group on the Home tab. Delete a record in a form by making the record active and then clicking the Delete button in the Records group on the Home tab.

Worldwide Enterprises

What You Will Do   Worldwide Enterprises has just signed two new distributors in New York and one company has signed with another movie distributing company. You will add two new records and delete one record from the US_Distributors table using a form.

Tutorial

Adding and Deleting Records in a Form

1 With **1-WEDistributors** open, double-click the *US_Distributors* form in the Forms group in the Navigation pane.

> The US_Distributors form opens with the first record in the US_Distributors table displayed in the form.

2 Click the New button [image] in the Records group on the Home tab.

> A blank form displays and the Record Navigation bar indicates you are editing record number 21. The New (blank) record and Next record buttons on the Record Navigation bar are dimmed.

3 Press the Tab key to move to the *CompanyName* field, since Access automatically assigns the next sequential number to the *DistID* field.

> The *DistID* field does not display a field value until you begin to type data in another field in the record.

4 Type Movie Emporium and then press the Tab key or the Enter key.

5 Type 203 West Houston Street and then press the Tab key or the Enter key.

> Use the same navigation methods you learned in Activity 1.5 to add the record to the table.

6 Type the remaining field values as shown at the right. Press the Tab key or the Enter key after typing the last field.

> When you press the Tab key or the Enter key after the *EmailAdd* field, a new form will appear in the work area.

In Brief

Add Record in Form View
1. Open form.
2. Click New (blank) record button in Navigation bar or click New button in Records group.
3. Type data in fields.

7 Type the following information in the appropriate fields in a new form for record 22:

Cinema Festival   212-555-9715
318 East 11th Street   212-555-9717
New York, NY 10003   cinemafest@ppi-edu.net

8 Click the First record button in the Record Navigation bar.

This displays the record for All Nite Cinemas.

9 Click in the first field in the *CompanyName* field column (*All Nite Cinemas*) and then click the Find button in the Find group on the Home tab.

10 At the Find and Replace dialog box with the Find tab selected, type Victory Cinemas and then click the Find Next button.

11 When the record for Victory Cinemas displays, click the Cancel button to close the Find and Replace dialog box.

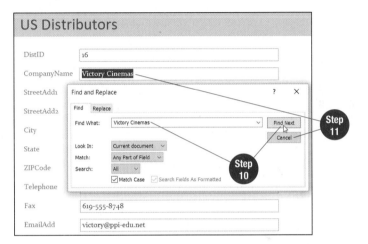

12 Click the Delete button arrow in the Records group on the Home tab and then click *Delete Record* at the drop-down list.

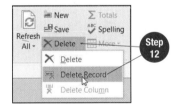

13 Click Yes to confirm the deletion.

14 Close the US_Distributors form.

15 Open the US_Distributors table and then view the two records added to and the one record deleted from the table using the form.

16 Close the US_Distributors table.

Check Your Work Compare your work to the model answer available in the online course.

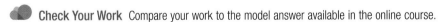

In Addition

Navigating in Form View Using the Keyboard

Navigate records in a form using the following keyboard commands:
- Press Page Down to display the next record.
- Press Page Up to display the previous record.

- Press Ctrl + End to move to the last field in the last record.
- Press Ctrl + Home to move to the first field in the first record.

Records in a table are displayed alphanumerically and sorted in ascending order by the primary key field values. To rearrange the order of records in a table, click in any entry in the field to sort by and then click the Ascending button or Descending button in the Sort & Filter group on the Home tab. Text is sorted from A to Z (ascending) or Z to A (descending), numbers (values) are sorted from lowest to highest (ascending) or highest to lowest (descending), and dates are sorted from earliest to latest (ascending) or latest to earliest (descending). To sort by more than one field column, select the field columns first and then click the Ascending or Descending button. Access sorts first by the leftmost field column in the selection and then by the next field column. Select adjacent field columns by positioning the mouse pointer on the header row of the first field column to be selected, click and hold down the left mouse button, drag to the header row of the last field column, and then release the mouse button. When a field is sorted, a sort icon displays in the field's header row. The sort icon displays as an up arrow if the sort is in ascending order and a down arrow if the sort is in descending order. Remove a sort by clicking the Remove Sort button on the Home tab. When a sort is removed, the records are sorted by the primary key field.

Worldwide Enterprises

What You Will Do   You will sort records in the *CompanyName* field, sort the records in the *ZIPCode* field, remove the sort, and then perform a multiple-field sort.

Tutorial

Sorting Records in a Table

① With **1-WEDistributors** open, open the US_Distributors table.

② Click in any record in the *CompanyName* field column.

③ Click the Ascending button in the Sort & Filter group on the Home tab.

> The records are rearranged to display the company names in ascending alphabetic order (A to Z) and a sort icon displays to the right of *CompanyName* in the header row.

④ Click the Descending button in the Sort & Filter group.

> The records are rearranged to display the companies in descending alphabetic order (Z to A). The sort icon changes from an up arrow to a down arrow.

⑤ Click in any record in the *ZIPCode* field column.

⑥ Click the Descending button in the Sort & Filter group.

⑦ Click the Remove Sort button in the Sort & Filter group.

> Removing the sort from the *ZIPCode* field sorts the records by the primary key field (the *DistID* field).

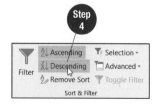

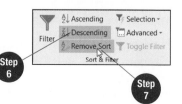

Sort Datasheet by Single Field
1. Open table.
2. Click in field in column by which to sort.
3. Click Ascending or Descending button.

Sort Datasheet by Multiple Fields
1. Open table.
2. Select columns from left to right in order of the sort.
3. Click Ascending or Descending button.

8 Position the mouse pointer in the *State* header row until the pointer changes to a down-pointing black arrow, click and hold down the left mouse button, drag to the right to select the *ZIPCode* field column, and then release the mouse button.

> The *State* and *ZIPCode* field columns are both selected.

Step 8

City	State	ZIPCode	Telephone
Louisville	KY	40201	502-555-4238
Arlington	TX	76013	817-555-2116
Columbia	SC	29201	803-555-3487
Cambridge	MA	02142	413-555-0981

9 Click the Ascending button in the Sort & Filter group.

10 Click in any field to deselect the two field columns.

> The records are sorted first by state and then by ZIP code within each state, as shown in Figure 1.4.

11 Close the US_Distributors table. Click Yes when prompted to save the design changes.

Figure 1.4 Records Sorted by State and Then by ZIP Code within Each State

US_Distributors

DistID	CompanyName	StreetAdd1	StreetAdd2	City	State	ZIPCode
7	LaVista Cinemas	111 Vista Road		Phoenix	AZ	86355-6014
10	Marquee Movies	1011 South Alameda Street		Los Angeles	CA	90045
15	Sunfest Cinemas	341 South Fourth Avenue		Tampa	FL	33562
8	Liberty Cinemas	P.O. Box 998	12011 Ruston Way	Atlanta	GA	73125
13	O'Shea Movies	59 Erie		Oak Park	IL	60302
11	Midtown Moviehouse	1033 Commercial Street		Emporia	KS	66801
1	All Nite Cinemas	2188 Third Street		Louisville	KY	40201
5	Eastown Movie House	P.O. Box 722	1 Concourse Avenue	Cambridge	MA	02142
20	Dockside Movies	P.O. Box 224	155 Central Avenue	Baltimore	MD	21203
21	Renaissance Cinemas	3599 Woodward Avenue		Detroit	MI	48211
6	Hillman Cinemas	55 Kemble Avenue		Baking Ridge	NJ	07920
23	Cinema Festival	318 East 11th Street		New York	NY	10003
22	Movie Emporium	203 West Houston Street		New York	NY	10014
19	Westview Movies	1112 Broadway		New York	NY	10119
17	Waterfront Cinemas	P.O. Box 3255		Buffalo	NY	14288
12	Mooretown Movies	P.O. Box 11	331 Metro Place	Dublin	OH	43107
14	Redwood Cinemas	P.O. Box 112F	336 Ninth Street	Portland	OR	97466-3359
18	Wellington 10	1203 Tenth Southwest		Philadelphia	PA	19178
4	Danforth Cinemas	P.O. Box 22	18 Pickens Street	Columbia	SC	29201
2	Century Cinemas	3687 Avenue K		Arlington	TX	76013
9	Mainstream Movies	P.O. Box 33	333 Evergreen Building	Seattle	WA	98220-2791

The ZIP codes are sorted in ascending order within each state.

Check Your Work Compare your work to the model answer available in the online course.

In Addition

Learning More about Sorting

When sorting records, consider the following alphanumeric rules:

- Numbers stored in fields that are not defined as numeric, such as social security numbers or telephone numbers, are sorted as characters (not numeric values). To sort them as numbers, all field values must be the same length.
- Records in which the selected field is empty are listed first.
- Numbers are sorted before letters.

Moving a Column

When sorting on more than one column, the columns must be adjacent. If necessary, columns can be moved to accommodate a multiple-field sort. To move a column, click the field name in the header row to select the column. Position the mouse pointer on the field name of the selected column, click and hold down the left mouse button, drag the thick black line that displays to the desired location, and then release the mouse button.

A filter is used to view only those records that meet specified criteria. The records that do not meet the filter criteria are hidden from view temporarily. Use a filter to view, edit, and/or print a subset of rows within the table. For example, a filter can be used to view only those records of distributors in one state. Apply a filter using the filter arrow at the right of the column heading or click the Filter button on the Home tab. Insert or remove check marks from the field entries to be filtered. Toggle between the data in the table and the filtered data using the Toggle Filter button on the Home tab and use the *Clear All Filters* option from the Advanced button drop-down list to clear all filters from the table. Records can be sorted by specific values using the *Text Filters* option at the filter drop-down list for fields containing text, the *Number Filters* option for fields that contain number values, or the *Date Filters* option for fields containing dates.

What You Will Do   You want to view a list of distributors in New York state and then further filter the list to display only those in the city of New York. In a second filter operation, you will view the distributors located in California and Georgia.

Filtering Records

1 With **1-WEDistributors** open, open the US_Distributors table.

2 Click in any field in the *State* field column.

3 Click the Filter button ▼ in the Sort & Filter group on the Home tab.

4 Click the *(Select All)* check box to remove the check marks from all of the check boxes.

> At the drop-down list, click the check boxes for those states that you do not wish to view to remove the check marks.

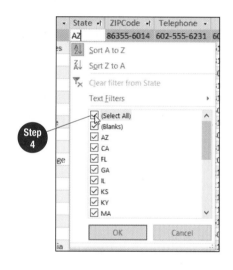

5 Scroll down the filter list box, click the check box next to *NY*, and then click OK.

> The filter list box closes and only four records remain, as shown below. Two icons display next to *State* indicating the field is both filtered and sorted and the message *Filtered* displays in the Record Navigation bar.

Only those records that meet the filter criterion, the *State* field, are displayed in Step 5.

6 Click the filter arrow at the right of *City* in the header row to display the filter drop-down list.

7 Click the check box next to *Buffalo* to remove the check mark and then click OK.

> The record for the company located in Buffalo is removed and only three records remain.

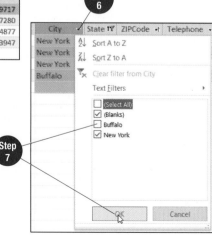

8 Click the Toggle Filter button ▼ in the Sort & Filter group.

> Click the Toggle Filter button to toggle between the data in the table and the filtered data. The message in the Record Navigation bar changes to *Unfiltered*.

9 Remove both filters from the table by clicking the Advanced button ▦ in the Sort & Filter group on the Home tab and then clicking *Clear All Filters*.

10 Save and then close the US_Distributors table.

11 Open the Orders table and then click in any field in the *Fee* field column.

12 Click the filter arrow at the right of *Fee* in the header row.

13 Point to *Number Filters* and then click *Less Than* at the side menu.

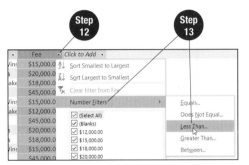

14 At the Custom Filter dialog box, type 15000.

> Typing *15000* in the Custom Filter dialog box specifies that you want to filter and display only those records with a fee amount less than or equal to $15,000.

15 Click OK.

> Six records display with a fee amount less than or equal to $15,000.

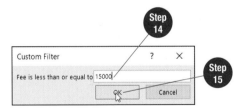

16 Close the Orders table and at the message that displays asking if you want to save the changes, click the No button.

In Addition

Filtering by Selection

Another method for filtering is to select specific text in a field and then click the Selection button in the Sort & Filter group on the Home tab. A drop-down list displays with filtering options such as *Equals* and *Does Not Equal*. The options vary depending on the data selected. For example, select *NY* in the *State* field in the US_Distributors table and the drop-down list displays the options *Equals "NY"*, *Does Not Equal "NY"*, *Contains "NY"*, and *Does Not Contain "NY"*. Click one of the options to filter the data.

To print a table in Datasheet view, click the File tab, click the *Print* option in the backstage area, and then click the *Quick Print* or *Print* option. To avoid wasting paper, use the *Print Preview* option at the Print backstage area to see how the table will appear on the page before printing the table. By default, Access prints a table on letter-size paper in portrait orientation. Change the paper size, orientation, or margins using buttons in the Page Size and Page Layout groups on the Print Preview tab.

Worldwide Enterprises

What You Will Do   Sam Vestering has requested a list of the US distributors. You will open the US_Distributors table, preview the printout, change the page orientation, change the left and right margins, and then print the table.

Tutorial

Previewing and
Printing a Table

1. With **1-WEDistributors** open, open the US_Distributors table.

2. Click the File tab, click the *Print* option, and then click the *Print Preview* option.

   The table displays in the Print Preview window as shown in Figure 1.5.

3. Move the mouse pointer (displays as a magnifying glass) over the top center of the table and then click the left mouse button.

   Clicking the left mouse button changes the display percentage to 100%. Access prints the table name at the top center and the current date at the top right of the page. At the bottom center, Access prints the word *Page* followed by the current page number.

4. Click the left mouse button again to display the current page in the Print Preview window.

5. Click the Next Page button on the Navigation bar at the bottom left of the Print Preview window.

   The US_Distributors table requires two pages to print with the default margins and orientation. In the next step, you will change to landscape orientation to see if all of the columns will fit on one page.

Figure 1.5 Print Preview Window

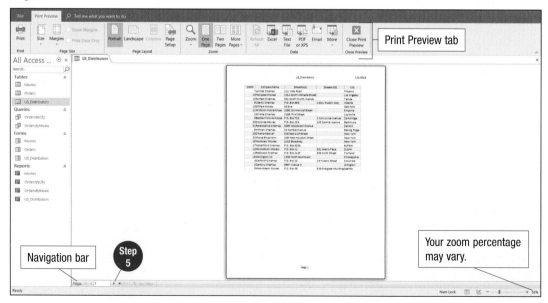

Display Table in Print Preview
1. Click File tab.
2. Click *Print* option.
3. Click *Print Preview* option.

Change to Landscape Orientation
1. Display table in Print Preview window.
2. Click Landscape button.
3. Click Close Print Preview button.

Change Margins
1. Display table in Print Preview window.
2. Click Margins button.
3. Click predefined margins option.
OR
1. Display table in Print Preview window.
2. Click Page Setup button.
3. Change margins settings.
4. Click OK.

6 Click the Landscape button in the Page Layout group on the Print Preview tab.

Landscape orientation rotates the printout to print wider than it is tall. Changing to landscape orientation allows more columns to fit on a page, but the US_Distributors table still needs two pages to print.

Step 6

7 Click the Margins button in the Page Size group and then click the *Narrow* option at the drop-down list.

The *Narrow* option changes all margins to 0.25 inches.

8 Click the Page Setup button in the Page Layout group.

9 At the Page Setup dialog box with the Print Options tab active, select *0.25* in the *Top* measurement box and then type 2.

10 Select *0.25* in the *Left* measurement box and then type .75.

11 Click OK.

12 Click the Print button in the Print group and then click OK at the Print dialog box.

In a few seconds, the table will print on the default printer installed on your computer. Even in landscape orientation with narrow margins, the table does not fit on one page. In Section 3, you will learn how to create a report for a table. Using a report, you can control the data layout on the page and which columns are printed.

13 Click the Close Print Preview button in the Close Preview group.

14 Close the US_Distributors table.

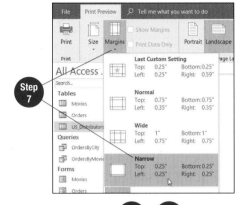

Step 7

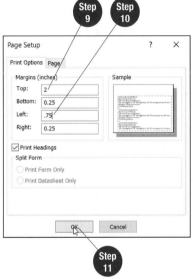

Step 9   Step 10

Step 11

Check Your Work Compare your work to the model answer available in the online course.

In Addition

Previewing Multiple Pages

Use buttons in the Zoom group on the Print Preview tab to view a specific number of pages in a multiple-page table. Click the Two Pages button to view the table with two pages side-by-side. Click the More Pages button and then choose *Four Pages*, *Eight Pages*, or *Twelve Pages* at the drop-down list. The US_Distributors table is shown at the right in the Two Pages view.

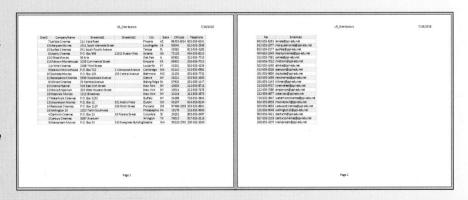

Activity 1.10   Using the Help and Tell Me Features; Hiding Columns in a Table

Microsoft provides help resources that contain information on Access features and commands. Press the F1 function key or click the Help tab and then click the Help button in the Help group to display the Help task pane. Or, open a dialog box, click the Help button in the upper right corner of the dialog box (displays as a question mark) and the Microsoft Office support website displays with information related to the dialog box. Click the Help button in the upper right corner of a backstage area and the Microsoft Office support website displays with information related to the specific backstage area. Access also includes the Tell Me feature, which provides information and guidance on how to complete a function. To use Tell Me, click in the *Tell Me* text box on the ribbon and then type the function. As text is typed in the *Tell Me* text box, a drop-down list displays with options that are refined as the text is typed. The drop-down list displays options for completing the function, displaying information on the function from sources on the Web, or displaying help information on the function. Hide a column in a table in Datasheet view if a field exists in the table that is not needed for data entry or editing purposes. Hiding the column provides more space in the work area, and hidden columns do not print. To hide a column, right-click the field name in the header row and then click *Hide Fields* at the shortcut menu.

What You Will Do   You will explore topics in the Help feature and use the Tell Me feature to change the font size of the text in the datasheet. Finally, you will reprint the formatted datasheet with two columns hidden that you decide you do not need on the printout.

Note: The following steps assume you are connected to the internet to access online resources.

Tutorial

Using the Help and Tell Me Features

1   With **1-WEDistributors** open, open the US_Distributors table.

2   Press the F1 function key to display the Help task pane.

> Find information in help resources by typing a search word or phrase and then pressing the Enter key or clicking the Search help button.

3   At the Help task pane, type keyboard shortcuts in the search text box and then press the Enter key.

> Keyboard shortcuts are helpful to know since they help you perform frequently used commands more quickly.

4   Click the Keyboard shortcuts for Access hyperlink.

5   Scroll down the articles in the Help task pane and then click the Edit text or data hyperlink.

6   Read the descriptions and keyboard shortcuts.

7   Close the Help task pane.

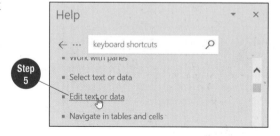

8 At the US_Distributors table, click in the *Tell Me* text box and then type font size.

A drop-down list displays with options such as *Font Size*, *Choose Page Size*, and *Font Color*.

9 Position the mouse pointer on the *Font Size* option at the drop-down list and then click *10* at the side menu.

The Tell Me feature guided you through the steps for changing font size of the text in the table.

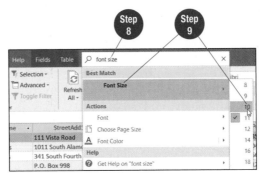

10 Right-click the *DistID* field name in the header row and then click *Hide Fields* at the shortcut menu.

11 Right-click the *EmailAdd* field name and then click *Hide Fields* at the shortcut menu.

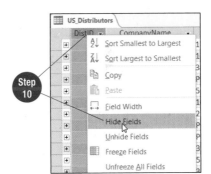

12 Click the File tab, click the *Print* option, and then click the *Print Preview* option.

13 Change to landscape orientation.

14 Print the table by clicking the Print button on the Print Preview tab and then clicking OK at the Print dialog box.

15 Close the Print Preview window and then close the US_Distributors table. Click Yes when prompted to save the layout changes.

16 Click the File tab and then click the *Close* option.

Check Your Work Compare your work to the model answer available in the online course.

In Addition

Accessing Help Information

The following options are available for displaying help information at the Help task pane:
- Press the F1 function key.
- Click the Help tab and then click the Help button.
- Type a function in the *Tell Me* text box, click the last option, and then click a resource at the side menu.

The following options are available for displaying help information at the Microsoft Office support website:
- Click the Help button in a dialog box.
- Click the Help button at a backstage area.

Features Summary

Feature	Ribbon Tab, Group	Button	Keyboard Shortcut
change font size	Home, Text Formatting		
change margins	Print Preview, Page Size OR Page Layout	OR to open Page Setup	
column width	Home, Records		
delete records	Home, Records		Delete
filter	Home, Sort & Filter		
find	Home, Find		Ctrl + F
Help	Help, Help		F1
landscape orientation	Print Preview, Page Layout		
Print backstage area	File, *Print*		
Print dialog box	File, *Print, Print*		Ctrl + P
Print Preview	File, *Print, Print Preview*		
replace	Home, Find	ab ↦ac	Ctrl + H
remove sort	Home, Sort & Filter	A Z	
save			Ctrl + S
sort ascending order	Home, Sort & Filter	A Z↓	
sort descending order	Home, Sort & Filter	Z A↓	
Tell Me feature	*Tell Me* text box		Alt + Q

Access

Creating Tables and Relationships

Data Files

Before beginning section work, copy the AccessS2 folder to your storage medium and then make AccessS2 the active folder.

Skills

- Create a new database
- Create a table in Datasheet view
- Assign data types to fields
- Create captions for fields
- Limit the number of characters allowed in a field
- Create a default value for a field
- Create and modify a table in Design view
- Set the primary key field for a table
- Verify data entry using a validation rule
- Restrict data entry using an input mask

- Set the format for displaying data
- Create a lookup list in a field
- Insert, move, and delete fields
- Apply formatting to data in a table
- Add a total row to a table
- Create, edit, and delete a one-to-many relationship
- Create and print a relationship report
- Create and edit a one-to-one relationship

Projects Overview

Worldwide Enterprises

Create and modify tables to store distributor contract information and employee benefit and review information; create relationships between tables. Review tables in an existing database and improve the table design.

Niagara Peninsula College

Create a table to store student grades for a course in the Theatre Arts Division.

The Waterfront Bistro

Create a new database to store contact information.

First Choice Travel

Create a new database to store employee expense claims. Modify a database on tour bookings.

Performance Threads

Modify and correct field properties in a costume inventory table to improve the design.

The online course includes additional training and assessment resources.

27

Activity 2.1 Creating a New Database; Creating a Table in Datasheet View

Tables form the basis for all other objects in a database and each table must be planned to adhere to database design principles. Creating a new table generally involves determining fields, assigning a data type to each field, designating the primary key field, and naming the table. This process is referred to as *defining the table structure*. As explained in Section 1, Access saves data automatically as records are entered into a datasheet. For this reason, to begin a new database, choose the Blank database template at the Access opening screen, assign the database a file name, and navigate to the storage location. Once the database has been created, a new blank table is presented in Datasheet view. Create the table and then enter records in the table.

Worldwide Enterprises

What You Will Do Riya Singh, a paralegal at Worldwide Enterprises, has asked you to create a new database in which distributor contracts can be maintained. You will create a new database file in which to store the contracts and then create the Contracts table. After the table is created, you will look at the table in a view that allows you to make changes to the table's structure.

Tutorial

Opening a Blank Database

Tutorial

Creating a Table in Datasheet View

1 Open Access and then click the *Blank database* template at the Access opening screen.

2 At the Blank database window, type 2-WEContracts in the *File Name* text box.

> One of the differences between Access and other programs is that the file name is assigned before any data is entered. Do not remove your storage medium while working in Access since the file will frequently have data written to it.

3 Click the Browse button 📁 at the right of the *File Name* text box.

4 At the File New Database dialog box, navigate to your AccessS2 folder and then click OK.

5 Click the Create button.

> A blank table appears in the work area with a tab labeled *Table1*. A column with the field name *ID* has been created automatically. The heading *Click to Add* appears at the top of the first blank column.

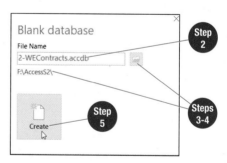

6 Review the table design shown in Figure 2.1. This is the table you will create in Steps 7–23. Refer to Table 2.1 for guidelines for naming fields. Note that in the Contracts table design the *ID* field that Access automatically creates will be used for the primary key field.

> Before you begin creating a table, refer to Table 2.2 on page 30 for information on field data types.

Figure 2.1 Fields for Contracts Table

Contracts		
Field Name	**Data Type**	**Sample Data**
*ID	AutoNumber	
ContractNo	Short Text	2021-034
Company	Short Text	West Coast Movies
ContactFN	Short Text	Jordan
ContactLN	Short Text	Daniels
Phone	Short Text	604-555-2886
Renewal	Yes/No	Yes
StartDate	Date & Time	July 1, 2021
EndDate	Date & Time	June 30, 2023
ShipFee	Currency	$500
Percentage	Number	10
Notes	Long Text	Jordan is an active member of the Western Film Association.

Table 2.1 Field Name Guidelines

A field name can be up to 64 characters and can include both letters and numbers. Some symbols are permitted, but others are excluded, so avoiding symbols is best. An exception is the underscore, which is often used as a word separator.

Do not use a space in a field name. Although a space is an accepted character, most database designers avoid using spaces in field names and object names. This practice facilitates easier management of the data with scripting or other database programming tools. Use compound words for field names or the underscore character as a word separator. For example, a field to hold a person's last name could be named *LastName*, *Last_Name*, or *LName*.

Abbreviate field names so that the names are as short as possible while still able to be readily understood. For example, a field such as *ContactLastName* could be abbreviated to *ContactLN* or a field such as *Telephone* to *Phone*. Shorter names are easier to manage and type into expressions.

7 Click the Short Text button AB in the Add & Delete group on the Table Tools Fields tab.

Access adds a new column to the datasheet with the Short Text data type assigned and selects the field name in the table header row for you to type the correct field name.

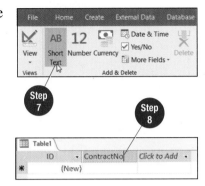

8 Type ContractNo and then press the Enter key.

Access moves to the *Click to Add* column and opens the data type drop-down list for the next new column.

9 Click *Short Text* at the drop-down list.

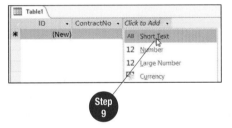

10 Type Company and then press the Enter key.

11 Click *Short Text* at the drop-down list, type ContactFN, and then press the Enter key.

12 Click *Short Text* at the drop-down list, type ContactLN, and then press the Enter key.

13 Click *Short Text* at the drop-down list, type Phone, and then press the Enter key.

14 Click *Yes/No* at the drop-down list.

The *Renewal* column will contain only one of two entries: *Yes* for those records that represent renewals of existing contracts or *No* if the record is a contract for a new distributor. When you change the data type to Yes/No, a check box is inserted in the column.

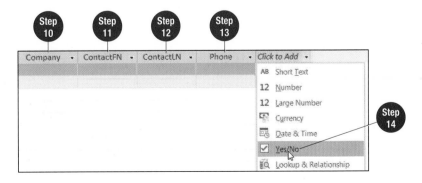

Table 2.2 Common Field Data Types in Datasheet View

Option	Description
Short Text	Alphanumeric data up to 255 characters in length, such as a name, address, or value such as a telephone number or social security number that is used as an identifier and not for calculating.
Number	Positive or negative values that can be used in calculations; do not use for values that will calculate monetary amounts (see *Currency*).
Currency	Values that involve money; Access will not round off during calculations.
Date & Time	Use this type to ensure dates and times are entered and sorted properly.
Yes/No	Data in the field will be either Yes or No, True or False, On or Off.
Lookup & Relationship	Can be used to enter data in the field from another existing table or display a list of values in a drop-down list for the user to choose from.
Long Text	Alphanumeric data up to 64,000 characters in length.
AutoNumber	Access automatically numbers each record sequentially (incrementing by 1) when you begin typing a new record.
Calculated Field	The contents of a calculated field are generated by Access from an expression you create; for example, a total cost could be calculated by adding a price plus a sales tax column.

15 Type Renewal and then press the Enter key.

16 Continue changing the data types and typing the field names for the remainder of the fields as shown in Figure 2.1 (on page 28).

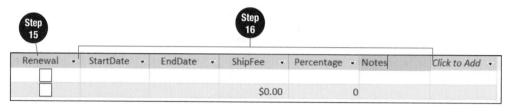

17 Press the Enter key after typing *Notes*. Click in any empty field below a column heading to remove the drop-down list and end the table.

18 If necessary, scroll to the left edge of the table. Click in the empty field below *ContractNo*, type 2021-034, and then press the Enter key or the Tab key to move to the next field.

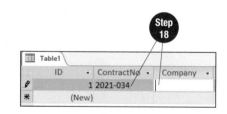

19 Continue typing the sample data for the remainder of the fields as shown in Figure 2.1 on page 28. At the first field in the *Renewal* column, click the check box or press the spacebar to insert a check mark in the box, indicating that the field entry is *Yes*.

Access converts dates you enter into the format *m/d/yyyy*. When entering dates, be careful to use the proper punctuation and spacing between the month, day, and year so that Access recognizes the entry as a valid date.

20 Click the Save button 🖫 on the Quick Access Toolbar.

In Brief

Create New Database
1. Start Access.
2. Click *Blank database* template.
3. Type database name.
4. Click Browse button.
5. Navigate to drive and/or folder.
6. Click OK.
7. Click Create button.

Create Table
1. Click data type in Add & Delete group on Table Tools Field s (Sp. Fields) tab.
2. Type field name and press Enter key.
3. Click field data type.
4. Type field name and press Enter key.
5. Repeat Steps 3–4 for remaining columns.
6. Click Save button.
7. Type table name.
8. Press Enter key or click OK.

21 At the Save As dialog box with *Table1* selected in the *Table Name* text box, type Contracts and then press the Enter key or click OK.

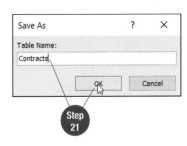

22 Click the View button ✓ in the Views group on the Home tab. (Do not click the button arrow.)

> The View button toggles between Design view and Datasheet view. In Design view, each row represents one field in the table and is used to define the field name, the field's data type, and an optional description. The *Field Properties* section in the lower half of the work area displays the properties for the active field. The properties will vary depending on the active field. A key icon displays in the field selector bar (blank column at the left of the field names) for the *ID* field, identifying it as the primary key field.

23 Compare the entries in the *Field Name* and *Data Type* columns for each field with those shown in Figure 2.1 on page 28. If necessary, correct a typing error in a field name by positioning the insertion point over the existing field name, clicking to open the field, and then inserting or deleting text as necessary. To change a data type, click in the *Data Type* column for the field, click the down arrow, and then click the correct data type at the drop-down list.

24 Click the View button ▦ in the Views group on the Table Tools Design tab. If you made changes while in Design view, you will be prompted to save the table when you switch views. If necessary, click the Yes button at any prompts that appear.

> Since you were in Design view, the View button changed to the Datasheet view button. Click the button to change back to Datasheet view.

25 Adjust the width of the columns to best fit the longest entries.

26 Print the table in landscape orientation and then close the table, saving the changes to the table layout.

Check Your Work Compare your work to the model answer available in the online course.

In Addition

Pinning/Unpinning a Database File

At the Access opening screen and the Open backstage area, Access displays the most recently opened databases. If a database is opened on a regular basis, consider pinning the database to the Recent list at the Access opening screen or to the *Recent* option list in the Open backstage area. To pin a database, hover the mouse over the database file name and then click the push pin at the right of the file name. Whether a database file is pinned at the Access opening screen or the Open backstage area, it displays in both. Unpin a database file by hovering the mouse over the database file name and then clicking the push pin at the right of the file.

Modifying Field Properties in Datasheet View

Each field has characteristics associated with it, called field properties. Field properties are used to control the behavior or interactivity of the field in database objects. Use buttons and options in the Properties group on the Table Tools Fields tab to modify field properties. Click the Name & Caption button and the Enter Field Properties dialog box displays with options for creating a descriptive title for fields. This is useful for providing a descriptive name if a field name is abbreviated in the table or to show spaces between words in a compound field name. Type a description in the *Description* text box and the description displays above the Status bar when entering records in the table. Use the *Field Size* option box to limit the number of characters that are allowed in a field entry. A field size of *4* for an *ID* field would prevent ID numbers longer than four characters from being stored in a record. If most records are likely to contain the same data, create a default entry to save typing and reduce errors. Click the Default Value button in the Properties group to display the Expression Builder dialog box with options for entering a default value.

Worldwide Enterprises

What You Will Do   You will modify the field properties of some of the fields in the Contracts table by providing a caption and description, specifying a field size, and/or providing a default value.

Tutorial

Modifying Field Properties in Datasheet View

1 With **2-WEContracts** open, open the Contracts table.

2 Click the *ContactFN* field name in the header row.

3 Click the Table Tools Fields tab.

4 Click the Name & Caption button in the Properties group.

5 At the Enter Field Properties dialog box, click in the *Caption* text box and then type First Name.

6 Click in the *Description* text box and then type Enter contact's first name.

7 Click OK to close the dialog box.

> The caption *First Name* now displays in the header row instead of *ContactFN*. The field name is still *ContactFN*, but the user-friendly caption displays in the header row.

8 Limit the field size for the contact's first name by clicking the number *255* in the *Field Size* option box in the Properties group, typing 20, and then pressing Enter.

9 At the message that displays indicating that the size of the field has been changed to a shorter size and asking if you want to continue, click Yes.

10 Click the *ContactLN* field name in the header row.

11 Click the Name & Caption button in the Properties group.

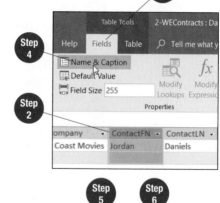

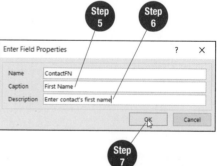

12 At the Enter Field Properties dialog box, click in the *Caption* text box and then type Last Name.

13 Click in the *Description* text box, type Enter contact's last name, and then click OK.

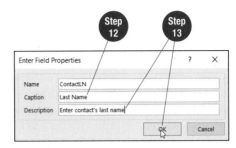

14 Click the number *255* in the *Field Size* option box, type 25, and then press the Enter key.

15 At the message that displays indicating that the size of the field has been changed to a shorter size and that some of the data may be lost, click the Yes button.

16 Click the *Renewal* field name in the header row.

17 Click the Default Value button in the Properties group.

18 At the Expression Builder dialog box, select the current default value *No* in the text box and then type Yes.

19 Click OK.

> With a default value of *Yes*, a check mark will automatically appear in the *Renewal* check box for new records entered in the table.

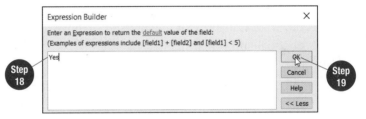

20 Print the table in landscape orientation and then close the table, saving the changes to the table layout.

21 Close **2-WEContracts**.

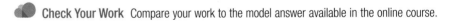

Check Your Work Compare your work to the model answer available in the online course.

In Addition

Changing an AutoNumber Data Type Field

Access automatically applies the AutoNumber data type to the first field in a table created in Datasheet view and assigns a unique number to each record. In many cases, letting Access automatically assign a number to a record is a good idea. Some situations may arise, however, when the unique value in the first field is something other than a number. Trying to change the AutoNumber data type in the first column by clicking one of the data type buttons in the Add & Delete group on the Table Tools Fields tab causes Access to create another field. To change the AutoNumber data type for the first field, click the *Data Type* option box arrow in the Formatting group on the Table Tools Fields tab and then click data type at the drop-down list.

As an alternative to creating a table in Datasheet view, a table can be created in Design view, where field properties can be set before entering data. When using Design view, Access does not add the *ID* field to the new table automatically. As mentioned previously, a table should have a primary key field, or a field used to store a unique value for each record. Examples of fields suitable for a primary key field are those that store an identification value such as an employee number, a part number, a vendor number, or a customer number. If, when designing a table, no data is suited to a primary key field, create a field labeled *ID* and set the data type to AutoNumber. When creating a table in Datasheet view, the Properties group on the Table Tools Fields tab contains buttons and an option for modifying or adding field properties. In Design view, the *Field Properties* section contains property boxes with options for adding or modifying field properties.

Worldwide Enterprises

What You Will Do   Rhonda Trask, human resources manager, has asked you to work in the employees database. Rhonda would like you to create a new table in the file in which to store the employee benefit plan information. You decide to create this table in Design view and add and modify field properties.

Tutorial

Creating a Table in Design View

1 Open **2-WEEmployees** and enable the content, if necessary.

> This database is similar to the database you worked with in the Skills Review for Section 1. This version of the database has additional records added.

Tutorial

Setting the Primary Key Field

2 Click the Create tab and then click the Table Design button ▦ in the Tables group.

3 With the insertion point positioned in the *Field Name* column in the first row, type EmployeeID and then press the Enter key or the Tab key to move to the next column.

Tutorial

Modifying Field Properties in Design View

4 With *Short Text* already entered in the *Data Type* column, change the default field size from *255* to *4* by double-clicking the number *255* in the *Field Size* property box in the *Field Properties* section and then typing 4.

> The *EmployeeID* field will contain numbers; however, leave the data type defined as *Short Text* since no calculations will be performed with employee numbers. Notice the information that displays at the bottom right corner of the Design window. Access displays information about the field in which the insertion point is positioned.

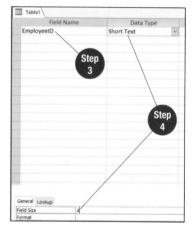

5 Click in the *Description* column at the right of *Short Text*, type Type the four-digit employee number, and then press the Enter key to move to the second row.

6 Type PensionPlan in the *Field Name* column in the second row and then press the Enter key.

7 Click the arrow at the right of the *Data Type* column and then click *Yes/No* at the drop-down list.

> In this field, the data is only one of two entries: *Yes* if the employee is enrolled in the pension plan or *No* if the employee is not.

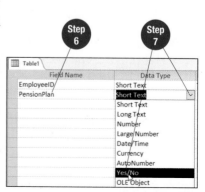

In Brief

Create Table in Design View
1. Click Create tab.
2. Click Table Design button.
3. Type field names, change data types, add descriptions, or modify other field properties.
4. Assign primary key field.
5. Click Save button.
6. Type table name.
7. Click OK.

Assign Primary Key Field
1. Open table in Design view.
2. Make primary key field active.
3. Click Primary Key button.
4. Save table.

8 Change the default value from *No* to *Yes* by double-clicking *No* in the *Default Value* property box in the *Field Properties* section and then typing Yes.

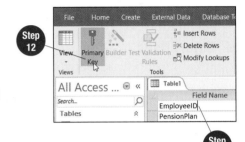

9 Click in the *Description* column at the right of *Yes/No*, type Click or press spacebar for Yes; leave empty for No, and then press the Enter key.

10 Enter the remaining field names, data types, and descriptions as shown in Figure 2.2 by completing steps similar to those in Steps 3–9. Change the default value for the *HealthPlan* field to *Yes*.

11 Click in the *EmployeeID* field name.

12 Click the Primary Key button 🔑 in the Tools group on the Table Tools Design tab.

> A key icon appears in the field selector bar at the left of *EmployeeID*, indicating the field is the primary key field.

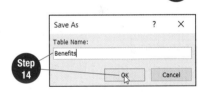

13 Click the Save button on the Quick Access Toolbar.

14 At the Save As dialog box, type Benefits in the *Table Name* text box and then press the Enter key or click OK.

15 Close the Benefits table.

Figure 2.2 Design View Table Entries

Field Name	Data Type	
EmployeeID	Short Text	Type the four-digit employee number
PensionPlan	Yes/No	Click or press spacebar for Yes; leave empty for No
DentalPlan	Yes/No	Click or press spacebar for Yes; leave empty for No
HealthPlan	Yes/No	Click or press spacebar for Yes; leave empty for No
Dependents	Number	Type the number of dependents related to this employee
LifeInsce	Currency	Type the amount of life insurance subscribed by this employee

Check Your Work Compare your work to the model answer available in the online course.

In Addition

Using the More Button on the Table Tools Fields tab in Datasheet View

Click the More Fields button to choose from a list of other field data types and fields that have predefined field properties. Scroll down the More Fields button drop-down list to the Quick Start category. Use options in this category to add a group of related fields in one step. For example, click *Address* to add five new fields with the captions: *Address*, *City*, *State Province*, *ZIP Postal*, and *Country Region*.

Activity 2.4   Applying Validation Rules

The accuracy of data in a table is extremely important. To control the data entered in a field and improve the accuracy of the data, apply a validation rule. For example, a validation rule can be applied that specifies amounts entered in a field must be less than a certain number. Apply a validation rule to a field in Datasheet view with the Validation button in the Field Validation group on the Table Tools Fields tab or with the *Validation Rule* property box in Design view. A validation message can be included that displays if the data entered in a field violates the validation rule. Use the Validation button in Datasheet view or the *Validation Text* property box in Design view to create a validation message.

What You Will Do   Worldwide Enterprises offers health benefits to a maximum of five dependents. You will use the Validation button on the Table Tools Fields tab to add a validation rule and message indicating that the number in the *Dependents* field must be less than six. The company offers life insurance in amounts up to a maximum of $200,000 per employee. You will use field properties in Design view to ensure that no amounts greater than $200,000 are entered in the *LifeIns* field.

Applying a Validation Rule in Datasheet View

Applying a Validation Rule in Design View

1. With **2-WEmployees** open, open the Benefits table in Datasheet view.

2. Click *Dependents* in the header row.

3. Click the Table Tools Fields tab.

4. Click the Validation button in the Field Validation group and then click the *Field Validation Rule* option.

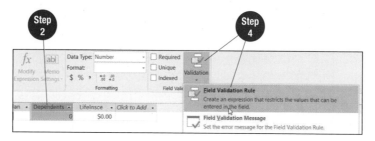

5. At the Expression Builder dialog box, type <6 in the text box and then click OK.

6. Click the Validation button and then click the *Field Validation Message* option.

7. Type Enter a number less than 6 in the text box in the Enter Validation Message dialog box and then click OK.

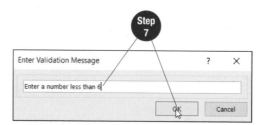

8. Click the View button in the Views group to switch to Design view.

9. Click *LifeInsce* in the *Field Name* column to display the associated field properties.

In Brief

Create Validation Rule
1. Open table in Design view.
2. Click in field row from which you want to create rule.
3. Click in *Validation Rule* property box.
4. Type statement.
5. Click in *Validation Text* property box.
6. Type error message.
7. Click Save.

10 Click in the *Validation Rule* property box, type <=200000, and then press the Enter key.

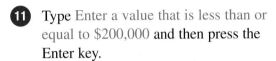

Pressing the Enter key after typing the validation rule moves the insertion point to the *Validation Text* property box. If an invalid syntax error displays, check your typing. Do not type a dollar symbol or comma in the validation rule statement. Also, make sure you have the correct less-than symbol (<) and equals sign (=) and that the order is <=.

11 Type Enter a value that is less than or equal to $200,000 and then press the Enter key.

12 Click the Save button.

13 Click the View button in the Views group to switch to Datasheet view.

14 Add the following record to the table (press the Enter key after typing *210000*):

EmployeeID	1003
PensionPlan	Yes
DentalPlan	Yes
HealthPlan	Yes
Dependents	2
LifeInsce	210000

When you enter *210000* into the *LifeInsce* field and then press the Enter key or the Tab key, Access displays an error message. The text in the error message is the text you entered in the *Validation Text* property box.

15 Click OK at the Microsoft Access error message.

16 Delete *210000*, type *200000*, and then press the Enter key.

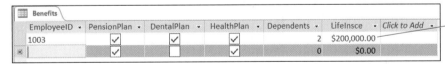

17 Close the Benefits table.

Check Your Work Compare your work to the model answer available in the online course.

In Addition

Understanding Other Types of Validation Rules

Validation rules should be created whenever possible to avoid data entry errors. The examples at the right illustrate various ways to use a validation rule to verify data.

Field Name	Validation Rule	Data Check
CustomerNo	>1000 And <1100	Limits customer numbers to 1001 through 1099.
CreditLimit	<=5000	Restricts credit limits to values of 5000 or less.
State	"CA"	Only the state of California is accepted.
Country	"CA" Or "US"	Only the United States or Canada is accepted.
OrderQty	>=25	Quantity ordered must be a minimum of 25.

Activity 2.5 Creating Input Masks; Formatting a Field

An input mask provides a pattern in a table or form indicating how data is to be entered into a field. For example, an input mask in a telephone number field that displays (___)___-____ indicates to the user that a three-digit area code is to be entered in front of all telephone numbers. Input masks ensure that data is entered consistently in tables. In addition to specifying the position and number of characters in a field, a mask can be created that restricts the data entered to digits, letters, or characters, and specifies whether or not each digit, letter, or character is required or optional. Create an input mask in Design view with the *Input Mask* property box. Use the *Format* property box to control how the data is displayed in the field after it has been entered.

Worldwide Enterprises

What You Will Do   You will create a new field in the Benefits table for pension plan eligibility dates and include an input mask and format property in the field.

Tutorial

Creating an Input Mask

1 With **2-WEEmployees** open, open the Benefits table in Design view.

2 Click in the *Field Name* column in the blank row below *LifeInsce*, type PensionDate, and then press the Enter key.

3 Change the data type to Date/Time and then press the Enter key.

4 Type Type date employee is eligible for pension plan in the format dd-mmm-yy (example: 31-Dec-21).

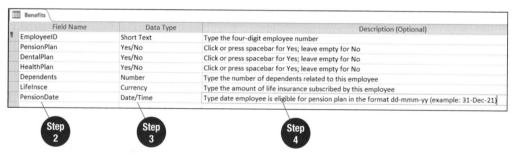

Field Name	Data Type	Description (Optional)
EmployeeID	Short Text	Type the four-digit employee number
PensionPlan	Yes/No	Click or press spacebar for Yes; leave empty for No
DentalPlan	Yes/No	Click or press spacebar for Yes; leave empty for No
HealthPlan	Yes/No	Click or press spacebar for Yes; leave empty for No
Dependents	Number	Type the number of dependents related to this employee
LifeInsce	Currency	Type the amount of life insurance subscribed by this employee
PensionDate	Date/Time	Type date employee is eligible for pension plan in the format dd-mmm-yy (example: 31-Dec-21)

Step 2   Step 3   Step 4

5 Click the Save button.

6 With the *PensionDate* field active, click in the *Input Mask* property box in the *Field Properties* section and then click the Build button **...** at the right side of the box.

7 Click *Medium Date* at the first Input Mask Wizard dialog box and then click the Next button.

The input masks that display in the list in the first dialog box are dependent on the data type for the field for which you are creating an input mask.

8 Click the Next button at the second Input Mask Wizard dialog box.

This dialog box shows the input mask code in the *Input Mask* text box and sets the placeholder character that displays in the field. The default placeholder is the underscore character.

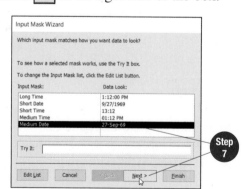

Step 7

In Brief

Use Input Mask Wizard
1. Open table in Design view.
2. Click in field row.
3. Click in *Input Mask* property box.
4. Click Build button.
5. Click input mask you want to create.
6. Click Next.
7. Select placeholder character.
8. Click Next.
9. Click Finish at the last wizard dialog box.
10. Click Save button.

9 Click the Finish button at the last Input Mask Wizard dialog box to complete the entry in the *Input Mask* property box and then press the Enter key.

10 Click in the *Format* property box.

The input mask controls how a date is entered into the field; however, by default, Access displays dates in the format *m/d/yyyy*. To avoid confusion, you will format the field to display the date in the same format that the input mask accepts the data.

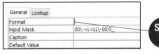

11 Click the arrow at the right side of the property box and then click *Medium Date* at the drop-down list.

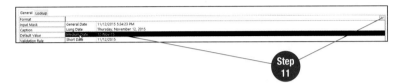

12 Click in the *Caption* property box and then type Pension Date.

13 Click the Save button and then switch to Datasheet view.

14 Click in the first field in the *Pension Date* column in the datasheet.

The input mask __-__-__ appears in the field.

15 Type 080721.

A beep sounds as you type every character after *08*. The only characters allowed after the first hyphen are letters. Notice the insertion point remains in the month section of the field.

16 With the insertion point positioned in the month section of the field, type jul21 and then press the Enter key.

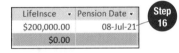

Notice that you did not have to type the hyphens or capitalize the first letter of the month. The greater than symbol (>) preceding *L* in the mask causes Access to convert the first character to uppercase.

17 Adjust the width of the *Pension Date* column to best fit the longest entry.

18 Close the Benefits table. Click Yes when prompted to save changes to the layout.

Check Your Work Compare your work to the model answer available in the online course.

In Addition

Input Mask Codes

The Input Mask Wizard is only available for fields with the Short Text or Date/Time data types. For fields with data types such as Number or Currency or for an input mask for which the wizard does not provide an option, create an input mask by entering the codes directly into the property box. At the right is a list of commonly used input mask codes.

Use	To restrict data entry to
0	digit, zero through nine, entry is required
9	digit or space, entry is not required
L	letter, A through Z, entry is required
?	letter, A through Z, entry is not required
>	all characters following are converted to uppercase
<	all characters following are converted to lowercase

Create a lookup field to restrict the data entered into the field to a specific list of items. The Lookup tab in the *Field Properties* section in Design view contains the options used to create a lookup field. Access includes the Lookup Wizard to facilitate entering the lookup settings.

Worldwide Enterprises

What You Will Do   You will create a new field in the Benefits table to store vacation entitlement for each employee. You want the field to display a drop-down list of vacation periods and restrict the field to accept only those entries that match items in the list.

Tutorial

Creating a Lookup Field

1 With **2-WEEmployees** open, open the Benefits table in Design view.

2 Click in the *Field Name* column in the blank row below *PensionDate*, type Vacation, and then press the Enter key.

3 Click the arrow at the right of the *Data Type* column and then click *Lookup Wizard* at the drop-down list.

4 At the first Lookup Wizard dialog box, click the *I will type in the values that I want* option and then click the Next button.

> If you press the Enter key by mistake and find yourself at the next step in the Lookup Wizard, click the Back button to return to the previous dialog box.

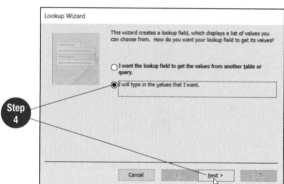

Step 4

5 Click in the blank row below *Col1*, type 1 week, and then press the Tab key.

6 Type 2 weeks and then press the Tab key.

7 Type 3 weeks and then press the Tab key.

8 Type 4 weeks and then click the Next button.

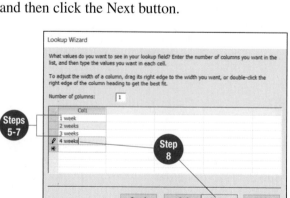

Steps 5-7

Step 8

9 Click the Finish button in the last Lookup Wizard dialog box to accept the default label *Vacation*. No entry is required in the *Description* column.

10 Click the Lookup tab in the *Field Properties* section and then view the entries made to each property by the Lookup Wizard.

11 Click in the *Limit To List* property box, click the arrow that appears, and then click *Yes*.

By changing the Limit To List property to *Yes*, you are further restricting the field to only those items in the drop-down list. If someone attempts to type an entry other than *1 week*, *2 weeks*, *3 weeks*, or *4 weeks*, Access will display an error message and will not store the data.

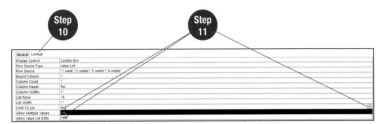

12 Click in the *Allow Value List Edits* property box, click the arrow that appears, and then click *No*.

You want to make sure that changes to the list that you created are not allowed by someone using the table or a form.

13 Click the Save button and then click the View button to switch to Datasheet view.

14 Click in the first field in the *Vacation* column, type 6 weeks, and then press the Enter key.

15 Click OK at the message informing you that the text entered isn't an item in the list.

16 Click *3 weeks* at the drop-down list and then press the Enter key.

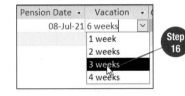

17 Display the table in Print Preview. Change to landscape orientation and then print the table.

18 Close the Benefits table.

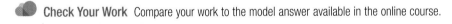 **Check Your Work** Compare your work to the model answer available in the online course.

In Addition

Looking Up Data from Another Table

Items in a drop-down list can also be generated by specifying an existing field in another table or query. To do this, click the Next button at the first Lookup Wizard dialog box to accept the default setting *I want the lookup field to get the values from another table or query*. In the remaining wizard dialog boxes, choose the table or query and the field to be used, choose the sort order for displaying the field values, adjust the column width for the lookup list, select the value to store, and assign a label to the column. Creating field entries using this method ensures that data is consistent between tables and eliminates duplicate typing of information that can lead to data errors. For example, in a database used to store employee information, one table could be used to enter employee numbers and then the remaining tables look up the employee number by scrolling a list of employee names.

Managing Fields; Formatting Data; Inserting a Total Row

Fields in Datasheet view can be inserted, moved, deleted, hidden, or frozen. Perform most field management tasks with buttons in the Records group on the Home tab. In Design view, fields can be inserted, moved, and deleted. Apply formatting to data in a table in Datasheet view with buttons and options in the Text Formatting group on the Home tab. Use the Totals button in the Records group on the Home tab to add a total row to a datasheet and then choose from a list of functions to find the sum, average, maximum, minimum, count, standard deviation, or variance result in a numeric column.

What You Will Do You will make changes to the structure of the Employees table in Datasheet view and Design view, apply formatting to the data in the table, and then add a total row and sum the amounts in the *AnnualSalary* column.

Tutorial
Managing Fields in Datasheet View

Tutorial
Managing Fields in Design View

Tutorial
Formatting Table Data

Tutorial
Inserting a Total Row

1 With **2-WEEmployees** open, open the Employees table in Datasheet view.

2 Click the *MiddleInitial* field column heading.

3 Click the Delete button ☒ in the Records group on the Home tab.

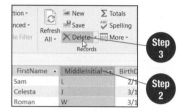

4 At the message asking you to confirm the deletion, click Yes.

5 Select the *Department* field column by clicking the *Department* column heading.

6 Position the mouse pointer in the *Department* column heading (make sure the mouse pointer displays as a white arrow); click and hold down the left mouse button, drag the thick, black, vertical line to the left so it is positioned between the *EmployeeID* and *LastName* field columns; and then release the mouse button.

When you release the mouse button, the *Department* field column is moved between the *EmployeeID* and *LastName* field columns.

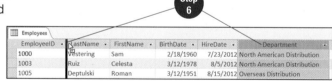

7 Click the View button to switch to Design view.

Even though the *Department* field column was moved in Datasheet view, the field remains in its original location in Design view.

8 Select the *AnnualSalary* field row by positioning the mouse pointer on the field selector bar at the left of *AnnualSalary* field row and then clicking the left mouse button.

9 Position the mouse pointer (white arrow) in the field selector bar for *AnnualSalary* field row; drag the black, horizontal line up so it displays between the *BirthDate* and *HireDate* field row; and then release the left mouse button.

The field moved in Design view remains in the original location in Datasheet view.

10 Click in a field in the *Department* row.

11 Click the Insert Rows button in the Tools group on the Table Tools Design tab.

A new blank row is inserted above the *Department* field row.

In Brief

Delete Field in Datasheet View
1. Click field name in header row.
2. Click Delete button.
3. Click Yes.

Delete Field in Design View
1. Select field row.
2. Click Delete Rows button.
3. Click Yes.

Insert Field in Design View
1. Make field active that will display after new field.
2. Click Insert Rows button.

Move Field in Datasheet View
1. Click field name in header row.
2. Drag vertical line to new position and then release button.

Move Field in Design View
1. Select field using field selector bar.
2. Drag horizontal line to new position and then release mouse button.

Insert Total Row
1. Display table in Datasheet view.
2. Click Totals button.

12 With the insertion point positioned in the *Field Name* column for the new row, type Supervisor and then press the Enter key. (Access will automatically assign the Short Text data type.)

13 Click the Save button and then click the View button to switch to Datasheet view.

14 Click the *Font* option box arrow in the Text Formatting group on the Home tab, scroll down the drop-down list, and then click *Candara*.

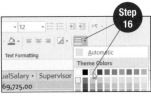

15 Click the *Font Size* option box arrow in the Text Formatting group and then click *12* at the drop-down list.

16 Click the Alternate Row Color button arrow in the Text Formatting group and then click the *Tan, Background 2* color option (third column, first row in the *Theme Colors* section).

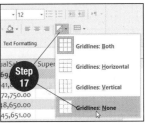

17 Click the Gridlines button in the Text Formatting group and then click *Gridlines: None* at the drop-down list.

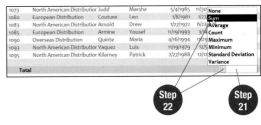

18 Select the *EmployeeID* field by clicking the *EmployeeID* column heading.

19 Click the Center button in the Text Formatting group.

20 Click the Totals button Σ in the Records group on the Home tab.

Access adds a row to the bottom of the table with the label *Total* at the left.

21 Click in the field in the *Total* row in the *AnnualSalary* column.

22 Click the arrow that appears at the left, click *Sum* at the drop-down list, and then click in any field in the table to deselect the total amount.

The sum *$1,163,808.00* appears at the bottom of the *AnnualSalary* field column.

23 Automatically adjust the width of the *Department* column to best fit the longest entry.

24 Display the table in Print Preview, change to landscape orientation, and then print the table.

25 Close the Print Preview window and then close the Employees table. Click Yes when prompted to save changes.

26 Click the File tab and then click the *Close* option.

Check Your Work Compare your work to the model answer available in the online course.

In Addition

Working with Wide Tables

When working in a table with many columns, scrolling right can make relating to the record in which changes need to be made difficult, since descriptor field columns such as *EmployeeID* or *LastName* may have scrolled off the screen. To alleviate this problem, freeze columns so they do not disappear when the datasheet is scrolled right. To do this, select the columns to freeze, click the More button in the Records group on the Home tab, and then click *Freeze Fields*.

Access is referred to as a *relational database management system*. A *relational database* is one in which tables have been joined. When two or more tables are joined to create a relationship, data can be looked up or reports can be created from multiple tables as if they were one table. Relationships help avoid data duplication. For example, in an employee database, the employee's first and last names would only need to appear in one table. In the remaining tables, only the employee identification number is needed. In most cases, tables are joined by a common field that exists in both tables. When two tables are joined in a relationship, one table is called the *primary table* and the other table is called the *related table*. One type of relationship is a *one-to-many relationship*, where one table in the relationship contains one unique record in the field used to join the tables while the other table can have several records with a matching field value in the joined field. In this type of relationship, the common field is the primary key field in the primary table and the field in the related table is the foreign key field. The referential integrity in a relationship can be enforced, which means that a value for the primary key field must first be entered in the primary table before it can be entered in the related table. Once referential integrity is enforced, other options become available for specifying that any change to the primary key field in the primary table is automatically updated in the related table and any record deleted in the primary table is automatically deleted in the related table. Establish a relationship between tables at the Relationships window. Display this window by clicking the Relationships button on the Database Tools tab.

Worldwide Enterprises

What You Will Do   You have been given a new employee database file that has additional records in the Benefits table and a new table that is used to store absence reports. You will create a one-to-many relationship between the Employees table and the Absences table using the common *EmployeeID* field. You will enforce referential integrity and cascade updated and deleted fields.

Tutorial

Creating a One-to-Many Relationship

1. Open **2-WEEmpRelationships** and enable the content, if necessary.

2. Open the Absences table and review the entries.

    This table is used to record employee absence reports. Notice that some employees have more than one record in the table.

3. Open the Benefits table and review the entries.

4. Open the Employees table and review the entries.

    Notice that the Employees table includes fields for the employees' names.

5. Click the Absences table tab at the top of the work area and notice that the names are not duplicated in the table. However, all three tables contain an *EmployeeID* field.

6. Close all three tables.

7. Click the Database Tools tab.

8. Click the Relationships button 🔳 in the Relationships group.

    The Relationships window opens in the work area.

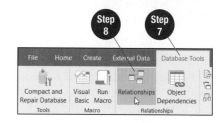

9 If necessary, click the Show Table button [image] in the Relationships group. (Skip this step if the Show Table dialog box displays in the Relationships window.)

10 At the Show Table dialog box with the Tables tab selected and *Absences* selected in the list box, click the Add button.

> A table field list box for the Absences table is added to the Relationships window. The Show Table dialog box remains open for you to add the next table.

11 Click *Benefits* in the list box and then click the Add button.

> Depending on the position of the Show Table dialog box in the Relationships window, the Benefits table field list box may be hidden behind the dialog box. If necessary, drag the Show Table dialog box title bar to move the dialog box out of the way.

12 Double-click *Employees* in the list box.

> Double-clicking a table name is another way to add a table field list box to the Relationships window.

13 Click the Close button in the Show Table dialog box.

> The three table field list boxes are side by side in the window. The *EmployeeID* field is the primary key field in the Employees table, but it is not the primary key field in the Absences table. This is because an employee can be absent many times; therefore, the *EmployeeID* field could not be a primary key field in the Absences table.

14 Position the mouse pointer on the bottom border of the Benefits table field list box until the pointer changes to an up-and-down arrow and then drag the bottom border down until the vertical scroll bar disappears.

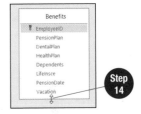

> This action expands the height of the table field list box so you can see all of the field names in the box.

15 Expand the height of the Employees table field list box to see all of the field names by completing a step similar to Step 14.

16 Position the mouse pointer on the Employees table field list box title bar and then drag the table field list box below the Benefits table field list box.

17 Move the Benefits table field list box to the right and then move the Employees table field list box up to fill in the space as shown below by dragging the table field list box title bars.

> You have been arranging the table field list boxes in the Relationships window to position the Employees table in the middle of the other two tables that will be joined. This layout will make it easier to create the relationship and understand the join lines that will appear.

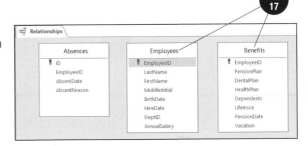

18 Position the mouse pointer on the *EmployeeID* field in the Employees table field list box, click and hold down the left mouse button, drag the pointer left to the *EmployeeID* field in the Absences table field list box, and then release the mouse button.

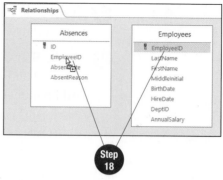

> A relationship is created by dragging the common field name from the primary table field list box to the related table field list box.

19 At the Edit Relationships dialog box, notice that *One-To-Many* displays in the *Relationship Type* section of the dialog box.

> Access determined the relationship type based on the common field that was used to join the tables. In the primary table (Employees), *EmployeeID* is the primary key field while in the related table (Absences), the *EmployeeID* field is not the primary key field. In the Absences table, the *EmployeeID* field is referred to as the *foreign key*.

20 Click the *Enforce Referential Integrity* check box to insert a check mark.

> *Referential integrity* means that Access will ensure that a record with the same employee number already exists in the primary table when a new record is added to the related table. This prevents what are known as *orphan records*—records in a related table for which no matching records are found in the primary table.

21 Click the *Cascade Update Related Fields* check box to insert a check mark.

> With a check mark in the *Cascade Update Related Fields* check box, make a change to a primary key field value and Access will automatically update the matching value in the related table.

22 Click the *Cascade Delete Related Records* check box to insert a check mark.

> With a check mark in the *Cascade Delete Related Records* check box, delete a record in the primary table and Access will delete any related records in the related table.

23 Click the Create button.

> A black line, called the *join line*, displays connecting the *EmployeeID* field in the Employees table field list box to the *EmployeeID* field in the Absences table field list box. Because you enforced referential integrity in the relationship, the join line displays with a *1* at the Employees table field list box and an infinity symbol at the Absences table field list box.

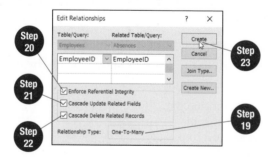

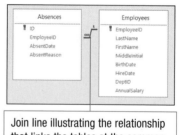

> Join line illustrating the relationship that links the tables at the common field in each table.

24 Click the Show Table button in the Relationships group.

25 At the Show Table dialog box, double-click *Departments* in the list box.

26 Click the Close button to close the Show Table dialog box.

In Brief

Create One-to-Many Relationship
1. Open Relationships window.
2. If necessary, add tables to window.
3. Close Show Table dialog box.
4. Drag common field name from primary table field list box to related table field list box.
5. Specify relationship options.
6. Click Create.
7. Click Save.

27 Position the mouse pointer on the *DeptID* field in the Departments table field list box, click and hold down the left mouse button, drag the pointer to the *DeptID* field in the Employees table field list box, and then release the mouse button.

28 Click the Create button.

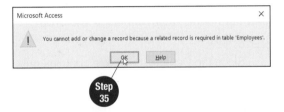

Since referential integrity was not enforced in this relationship, the black join line does not display with a 1 at one end and the infinity symbol at the other.

29 Click the Save button.

30 Click the Close button ☒ on the Relationship Tools Design tab to close the Relationships window.

31 Open the Absences table.

32 Click in the blank row at the bottom of the datasheet in the *EmployeeID* field column, type 1099, and then press the Enter key.

33 Type March 3, 2021 in the *Absent Date* field column and then press the Enter key.

34 With *Sick Day* the default value in the *Absent Reason* field column, press the Enter key to accept the entry.

Access displays an error message indicating you cannot add or change a record because a related record is required in the Employees table.

35 Click OK at the Microsoft Access message box.

36 Close the Absences table. Click OK at the Microsoft Access error message that appears for the second time.

37 Click Yes at the second error message box to close the table and confirm that the data changes will be lost.

🔵 **Check Your Work** Compare your work to the model answer available in the online course.

In Addition

Clearing the Relationships Window Layout

To clear the layout of the table field list boxes in the Relationships window and to remove relationships created between tables, click the Clear Layout button in the Tools group on the Relationship Tools Design tab. At the message indicating the the layout will be cleared and asking if you want to continue, click the Yes button.

A relationship between tables can be edited with options at the Edit Relationships dialog box. Display this dialog box at the Relationships window by clicking the Edit Relationships button on the Relationship Tools tab. Relationships can be deleted if the database is redesigned and the existing relationship no longer applies. A relationship also may need to be deleted in order to make a structural change to a table. In this case, the relationship can be deleted, the change made, and then the relationship re-created. Once all relationships have been created in a database, printing a hard copy of the relationship report to file away for future reference is a good idea. This documentation is a quick reference of all the table names, fields within each table, and relationships between the tables. Should a relationship need to be re-created, this documentation will be of assistance.

Worldwide Enterprises

What You Will Do   You will edit the relationship between the Departments table and Employees table, delete a relationship, and then create and then print a relationship report.

Editing and Deleting
a Relationship

Creating a Relationship
Report

1 With **2-WEEmpRelationships** open, open the Relationships window by clicking the Database Tools tab and then clicking the Relationships button in the Relationships group.

2 Click the Edit Relationships button in the Tools group.

3 At the Edit Relationships dialog box, click the *Table/Query* option box arrow and then click *Departments* at the drop-down list.

> Since the Departments table contains only one relationship, *Employees* automatically displays in the *Related Table/Query* option box and *DeptID* displays below each table name.

4 Click the *Enforce Referential Integrity* check box, the *Cascade Update Related Fields* check box, and the *Cascade Delete Related Records* check box to insert check marks.

5 Click OK.

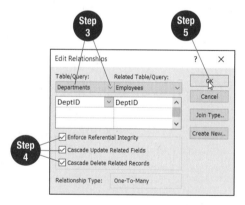

6 Right-click the black join line between the Absences table field list box and the Employees table field list box.

7 Click *Delete* at the shortcut menu.

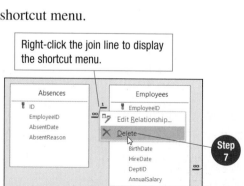

Right-click the join line to display the shortcut menu.

8 Click Yes at the Microsoft Access message box asking if you are sure you want to permanently delete the relationship.

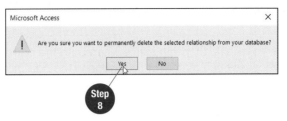

Step 8

9 Click the Relationship Report button 🖼 in the Tools group on the Relationship Tools Design tab.

Access generates the report and displays it in Print Preview in a new tab in the work area.

10 Click the Print button 🖨 in the Print group.

Step 10

11 At the Print dialog box, click OK.

12 Click the Close button at the top right of the work area to close the report.

13 Click Yes at the Microsoft Access message box asking if you want to save the report.

14 At the Save As dialog box with the existing report name already selected in the *Report Name* text box, type Relationships and then click OK or press the Enter key.

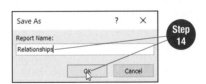

Step 14

15 Close the Relationships window.

Check Your Work Compare your work to the model answer available in the online course.

In Addition

Displaying Records in a Subdatasheet

When two tables are joined, related records within a subdatasheet can be viewed. A subdatasheet is a datasheet within a datasheet. When the primary table is opened in Datasheet view, a column of plus symbols (referred to as *expand indicators*) displays between the record selector bar and the first column. To view related records, click the expand indicator next to the specific record. A subdatasheet opens, similar to the one shown below. The plus symbol changes to a minus symbol when a record has been expanded. Click the minus symbol (referred to as a *collapse indicator*) to close the subdatasheet.

Click to collapse the record and remove the subdatasheet.

subdatasheet

Activity 2.10   Creating a One-to-One Relationship

A one-to-one relationship exists when both the primary table and the related table contain only one record with a matching field value in the common field. In this relationship, the common field used to join the tables is the primary key field in each table. For example, the Employees table would contain only one record for each employee. The Benefits table would also contain only one record for each employee. If these tables are joined on the common *EmployeeID* field, a one-to-one relationship would be created. In this type of relationship, consider the primary table to be the table with the fields (such as the employee names) that describe the identity of each employee number (the person for whom the employee number was created).

Worldwide Enterprises

Tutorial

Creating a One-to-One Relationship

What You Will Do   You will create a one-to-one relationship between the Employees and the Benefits tables and then edit the relationship after it has been created.

1. With **2-WEEmpRelationships** open, open the Relationships window.

2. Notice the *EmployeeID* field is the primary key field in both the Employees table field list box and the Benefits table field list box.

3. Position the mouse pointer on the *EmployeeID* field in the Employees table field list box, click and hold down the left mouse button, drag the pointer to the *EmployeeID* field in the Benefits table field list box, and then release the mouse button.

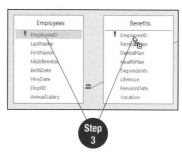
Step 3

> Access determined the relationship type as one-to-one since the common field that was used to join the two table field list boxes is the primary key field in each table. In both tables, only one record can exist for each unique employee number.

4. Click the Create button.

> The black join line connecting the two *EmployeeID* fields appears between the two table field list boxes in the Relationships window. The join line does not show a *1* at each end because referential integrity was not enforced.

5. Click the Save button.

6. Click the Close button on the Relationship Tools Design tab to close the Relationships window.

> In the next steps, you will add a record to the Benefits table to illustrate why referential integrity is a good idea to ensure primary tables are updated first.

7. Open the Benefits table and then add the following record to the table:

EmployeeID	1100
Pension Plan	Yes
Dental Plan	Yes
Health Plan	Yes
Dependents	5
Life Insurance	200000
Pension Date	31-Dec-21
Vacation	1 week

⊞ 1095				1	$200,000.00	31-Dec-21 1 week
⊞ 1100	✓	✓	✓	5	$200,000.00	31-Dec-21 1 week

Step 7

In Brief

Create a One-to-One Relationship
1. Open Relationships window.
2. If necessary, add tables to window.
3. Close Show Table dialog box.
4. Drag common field name from primary table field list box to related table field list box.
5. Specify relationship options.
6. Click Create.
7. Click Save.

8 Open the Employees table. If necessary, scroll down to view the last record in the table. No employee record exists in the table with an employee ID of 1100.

> Since referential integrity was not turned on when the relationship was created between the Employees and the Benefits tables, you were able to add a record to the related table for which no matching record is found in the primary table. Although you could easily add the matching record to the Employees table afterwards, establishing the employee's record in the first primary table is considered good practice.

9 Close the Employees table.

10 Delete the record you added at Step 7, the record with *1100* in the *EmployeeID* field in the Benefits table, and then close the table.

11 Open the Relationships window.

12 Right-click the black join line between the Employees table field list box and the Benefits table field list box and then click *Edit Relationship* at the shortcut menu.

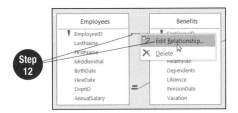

13 Click the *Enforce Referential Integrity* check box at the Edit Relationships dialog box and then click OK.

> With referential integrity turned on, the join line displays a *1* at each end.

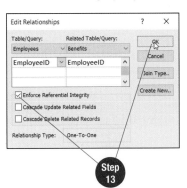

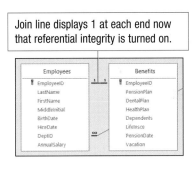

Join line displays 1 at each end now that referential integrity is turned on.

14 Close the Relationships window.

15 Close **2-WEEmpRelationships**.

🔵 **Check Your Work** Compare your work to the model answer available in the online course.

Features Summary

Feature	Ribbon Tab, Group	Button
alternate row color	Home, Text Formatting	
create table	Create, Tables	
create table in Design view	Create, Tables	
Datasheet view	Home, Views	
delete fields	Table Tools Design, Tools	
edit relationships	Relationship Tools Design, Tools	
font	Home, Text Formatting	
font size	Home, Text Formatting	
gridlines	Home, Text Formatting	
insert fields	Table Tools Design, Tools	
insert totals	Home, Records	
primary key field	Table Tools Design, Tools	
relationship report	Relationship Tools Design, Tools	
relationships	Database Tools, Relationships	

Access

Creating Queries, Forms, and Reports

Data Files

Before beginning section work, copy the AccessS3 folder to your storage medium and then make AccessS3 the active folder.

Skills

- Create a query using the Simple Query Wizard
- Create a query in Design view using one table
- Create a query in Design view using multiple tables
- Add criteria statements to a query
- Sort and hide columns in a query results datasheet
- Design a query with an *And* criteria statement and design a query with an *Or* criteria statement

- Perform calculations in a query
- Create and format a form
- Add an existing field to a form
- Manage control objects in a form
- Create and format a report
- Manage control objects in a report

Projects Overview

Create queries to produce custom employee lists, add criteria, and calculate pension contributions and monthly salaries; create and modify forms to facilitate data entry and viewing of records; create and modify reports to produce custom printouts of data.

Create a query, and create and print a report that lists all costumes rented in a particular month; create and modify a form for browsing the costume inventory and entering new records; continue design work on a new database for custom costume activities by creating a form and a report.

NIAGARA PENINSULA COLLEGE

Create and print a query to extract records of students who achieved A+ in all of their courses.

The online course includes additional training and assessment resources.

The Waterfront BISTRO

Create queries and design a report for the catering events database to extract event information for a banquet room, extract all events booked in a particular month, and calculate the estimated revenue from the catering events.

53

A *query* is an Access object designed to extract data from one or more tables. Usually a query is created to select records that answer a question. For example, a question such as *Which employees are enrolled in the Pension Plan?* could be answered with a query. The query would be designed to select records for those employees with a *Yes* in the *PensionPlan* field. Query results display in a datasheet that pulls the data from existing tables. A query can be created to serve a variety of purposes, from very simple field selections to complex conditional statements or calculations. In its simplest form, a query may be used to display or print selected fields from two tables. Access includes the Simple Query Wizard to facilitate creating a query.

Worldwide Enterprises

What You Will Do   Using the Simple Query Wizard, you will generate a list of each employee's first and last names and benefit selections. This will allow you to print a list by selecting fields from two tables.

Tutorial

Creating a Query Using the Simple Query Wizard

1 Open **3-WEEmployees** and enable the content, if necessary.

2 Click *Employees* in the Tables group in the Navigation pane and then click the Create tab.

3 Click the Query Wizard button ▦ in the Queries group.

4 At the New Query dialog box, with *Simple Query Wizard* already selected in the list box, click OK.

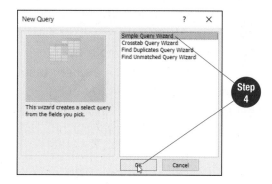
Step 4

5 At the first Simple Query Wizard dialog box, with *Table: Employees* selected in the *Tables/Queries* option box and *EmployeeID* selected in the *Available Fields* list box, click the One Field button ⟩ to move *EmployeeID* to the *Selected Fields* list box.

6 With *LastName* now selected in the *Available Fields* list box, click the One Field button to move *LastName* to the *Selected Fields* list box.

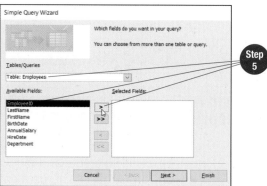
Step 5

7 Click the One Field button to move *FirstName* to the *Selected Fields* list box.

8 Click the the *Tables/Queries* option box arrow and then click *Table: Benefits* at the drop-down list.

> The list of fields in the *Available Fields* list box changes to display the field names from the Benefits table.

9 Double-click *PensionPlan* in the *Available Fields* list box.

> Double-clicking a field name is another way to move a field to the *Selected Fields* list box.

In Brief

Create Query Using Simple Query Wizard
1. Click Create tab.
2. Click Query Wizard button.
3. Click OK.
4. Choose table(s) and field(s) to include in query.
5. Click Next.
6. Choose *Detail* or *Summary* query.
7. Click Next.
8. Type title for query.
9. Click Finish.

10. Double-click the following fields in the *Available Fields* list box to move them to the *Selected Fields* list box:

    DentalPlan
    HealthPlan
    Vacation

11. Click the Next button.

12. Click the Next button at the second Simple Query Wizard dialog box to accept *Detail (shows every field of every record)* in the *Would you like a detail or summary query?* section.

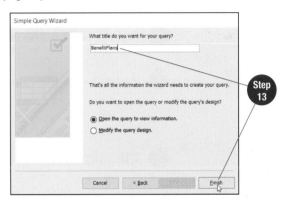

13. At the third Simple Query Wizard dialog box, select the current text in the *What title do you want for your query?* text box, type BenefitPlans, and then click the Finish button.

    View the query results datasheet shown in Figure 3.1. A query results datasheet can be sorted, edited, or formatted in a manner similar to a table. Data displayed in query results is not stored as a separate entity—the query is simply another interface for viewing and editing data in the associated table(s). Each time a saved query is opened, Access dynamically updates the query results by running the query.

14. Display the query in Print Preview, change to landscape orientation, and then print the query.

15. Close the Print Preview window and then close the BenefitPlans query.

Figure 3.1 BenefitPlans Query Results Datasheet

EmployeeID	LastName	FirstName	Pension Plan	Dental Plan	Health Plan	Vacation
1000	Vestering	Sam	☑	☑	☑	4 weeks
1003	Ruiz	Celesta	☑	☑	☑	3 weeks
1005	Deptulski	Roman	☑	☑	☐	3 weeks
1013	Chippewa	Gregg	☑	☑	☑	3 weeks
1015	Brewer	Lyle	☐	☑	☑	4 weeks
1020	Doxtator	Angela	☑	☑	☐	3 weeks
1023	Bulinkski	Aleksy	☐	☐	☐	3 weeks
1025	Biliski	Jorge	☑	☑	☑	4 weeks
1030	Hicks	Thom	☐	☐	☐	2 weeks
1033	Titov	Dina	☑	☑	☑	3 weeks
1040	Lafreniere	Guy	☑	☑	☑	3 weeks
1043	Morano	Enzo	☑	☑	☑	3 weeks
1045	Yiu	Terry	☐	☐	☐	2 weeks
1050	Zakowski	Carl	☑	☐	☑	2 weeks
1053	O'Connor	Shauna	☑	☑	☑	3 weeks
1060	McKnight	Donald	☑	☑	☑	3 weeks
1063	McPhee	Charlotte	☑	☐	☑	2 weeks
1065	Liszniewski	Norm	☑	☑	☑	2 weeks
1073	Judd	Marsha	☑	☑	☑	2 weeks
1080	Couture	Leo	☑	☑	☐	1 week
1083	Arnold	Drew	☑	☐	☑	1 week
1085	Armine	Yousef	☐	☑	☑	1 week
1090	Quinte	Maria	☑	☐	☑	1 week
1093	Vaquez	Luis	☑	☑	☐	1 week
1095	Kilarney	Patrick	☐	☐	☐	1 week

 Check Your Work Compare your work to the model answer available in the online course.

Section 2 provided information on working with tables in Design view to define or modify the table structure. Similarly, Design view can be used to create a query. Designing a query consists of identifying the table or tables containing the data to be extracted, the field or fields from which the data will be drawn, and the critera for selecting the data. After designing the query, display the records matching the criteria by switching to Datasheet view or by clicking the Run button.

Worldwide Enterprises

What You Will Do   Rhonda Trask, human resources manager, has asked for a list of employees with their annual salaries and hire dates. You will produce the list by creating a query to obtain the required fields from the Employees table.

Tutorial

Creating a Query in Design View

1 With **3-WEEmployees** open and the Create tab active, click the Query Design button [icon] in the Queries group.

2 At the Show Table dialog box with the Tables tab selected, double-click *Employees*.

> A table field list box for the Employees table is added to the query. The first step in building a query in Design view is to add a table field list box for each table from which records will be selected.

3 Click the Close button to close the Show Table dialog box.

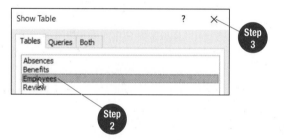

4 Double-click the *EmployeeID* field in the Employees table field list box.

> The blank columns at the bottom represent the columns in the query results datasheet and are referred to as the *query design grid*. You place the field names in the columns in the order in which you want the fields displayed in the query results datasheet. Double-clicking a field name adds the field to the next available column. In Steps 5 and 6, you will practice two other methods of adding fields to the query design grid.

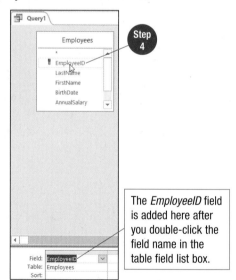

The *EmployeeID* field is added here after you double-click the field name in the table field list box.

5 Position the mouse pointer on the *FirstName* field in the Employees table field list box, click and hold down the left mouse button, drag to the second field in the *Field* row of the query design grid, and then release the mouse button.

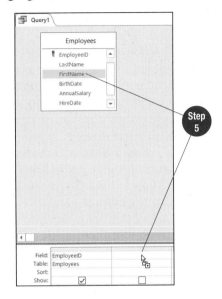

6 Click in the third field in the *Field* row of the query design grid, click the arrow that appears, and then click *LastName* at the drop-down list.

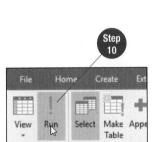

7 Using any of the three methods in Steps 4–6, add the *AnnualSalary* and *HireDate* fields from the Employees table field list box to the query design grid.

8 Click the Save button 💾 on the Quick Access Toolbar.

9 At the Save As dialog box, type SalaryList in the *Query Name* text box and then press the Enter key or click OK.

10 Click the Run button 🔲 in the Results group on the Query Tools Design tab.

> A query stores instructions on how to select data. The Run command instructs Access to carry out the instructions and display the results.

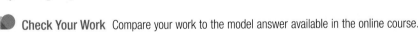

11 Print the query.

12 Close the SalaryList query.

Check Your Work Compare your work to the model answer available in the online course.

In Addition

Understanding Action Queries

In the last activity and in this activity, select queries were created that displayed selected fields from tables. Another type of query, called an *action query*, makes changes to a group of records. Four types of action queries are available in Access: delete, update, append, and make-table. A delete query deletes records. An update query makes global changes to a field. An append query adds a group of records from one table to the end of another table. A make-table query creates a new table from all or part of the data in existing tables.

Often a query is used to select records from more than one table. In Activity 3.1, a query results datasheet displayed records from the Employees table and the Benefits table. In Design view, multiple tables are added to the query at the Show Table dialog box. Once a table field list box has been added for each table, the fields are added to the query design grid in the desired order using any one of the three methods learned in the last activity.

What You Will Do   Rhonda Trask has asked for a list of employees, along with their hire and review dates. This data is stored in two different tables. You will produce the list by creating a query to obtain the required fields from each table to generate the list.

Creating a Query in Design View Using Multiple Tables

1 With **3-WEEmployees** open and the Create tab active, click the Query Design button.

2 At the Show Table dialog box with the Tables tab selected, double-click *Employees*.

3 Double-click *Review* and then click the Close button to close the Show Table dialog box.

A black join line with *1* at each end of the line between the Employees table field list box and the Review table field list box illustrates the one-to-one relationship that has been defined between the two tables.

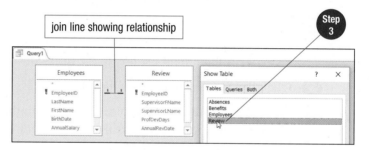

4 Double-click the *EmployeeID* field in the Employees table field list box.

5 Double-click the *FirstName* field in the Employees table field list box.

6 Double-click the *LastName* field in the Employees table field list box.

7 Double-click the *SupervisorFName* field and then double-click the *SupervisorLName* field in the Review table field list box to add the fields from the second table to the query design grid.

8 Double-click the *HireDate* field in the Employees table field list box.

You can add fields in any order from either table to the query design grid.

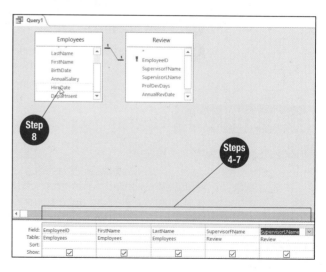

⑨ Double-click the *AnnualRevDate* field in the Review table field list box.

⑩ Look at the table names in the *Table* row in the query design grid. The table with which each field is associated is displayed.

Step 10

⑪ Click the Save button on the Quick Access Toolbar.

⑫ At the Save As dialog box, type ReviewList in the *Query Name* text box and then press the Enter key or click OK.

Step 12

⑬ Click the Run button in the Results group on the Query Tools Design tab.

View the query results in Figure 3.2.

⑭ Display the datasheet in Print Preview, change to landscape orientation, and then print the datasheet.

⑮ Close the Print Preview window and then close the ReviewList query.

Figure 3.2 ReviewList Query Results Datasheet

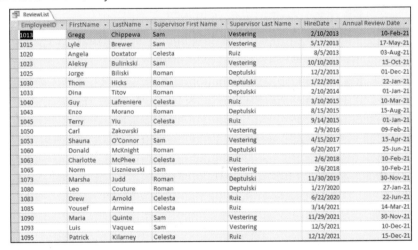

EmployeeID	FirstName	LastName	Supervisor First Name	Supervisor Last Name	HireDate	Annual Review Date
1013	Gregg	Chippewa	Sam	Vestering	2/10/2013	10-Feb-21
1015	Lyle	Brewer	Sam	Vestering	5/17/2013	17-May-21
1020	Angela	Doxtator	Celesta	Ruiz	8/5/2013	03-Aug-21
1023	Aleksy	Bulinkski	Sam	Vestering	10/10/2013	15-Oct-21
1025	Jorge	Biliski	Roman	Deptulski	12/2/2013	01-Dec-21
1030	Thom	Hicks	Roman	Deptulski	1/22/2014	22-Jan-21
1033	Dina	Titov	Roman	Deptulski	2/10/2014	01-Jan-21
1040	Guy	Lafreniere	Celesta	Ruiz	3/10/2015	10-Mar-21
1043	Enzo	Morano	Roman	Deptulski	8/15/2015	15-Aug-21
1045	Terry	Yiu	Celesta	Ruiz	9/14/2015	01-Jan-21
1050	Carl	Zakowski	Sam	Vestering	2/9/2016	09-Feb-21
1053	Shauna	O'Connor	Sam	Vestering	4/15/2017	15-Apr-21
1060	Donald	McKnight	Roman	Deptulski	6/20/2017	25-Jun-21
1063	Charlotte	McPhee	Celesta	Ruiz	2/6/2018	10-Feb-21
1065	Norm	Liszniewski	Sam	Vestering	2/6/2018	10-Feb-21
1073	Marsha	Judd	Roman	Deptulski	11/30/2019	30-Nov-21
1080	Leo	Couture	Roman	Deptulski	1/27/2020	27-Jan-21
1083	Drew	Arnold	Celesta	Ruiz	6/22/2020	22-Jun-21
1085	Yousef	Armine	Celesta	Ruiz	3/14/2021	14-Mar-21
1090	Maria	Quinte	Sam	Vestering	11/29/2021	30-Nov-21
1093	Luis	Vaquez	Sam	Vestering	12/5/2021	10-Dec-21
1095	Patrick	Kilarney	Celesta	Ruiz	12/12/2021	15-Dec-21

Check Your Work Compare your work to the model answer available in the online course.

In Addition

Learning More about Adding Tables to the Query Design Grid

If, after closing the Show Table dialog box, another table field list box needs to be added to the query design grid, click the Show Table button in the Query Setup group on the Query Tools Design tab to redisplay the dialog box.

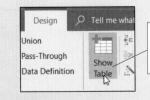

Click the Show Table button to add a table field list box to the query design grid.

In the previous queries, all records from the tables were displayed. Adding a criterion statement to the query design grid will cause Access to display only those records that meet the criterion. For example, a list of employees who are entitled to four weeks of vacation can be generated. Extracting specific records from tables is where the true power in creating queries is found. Use the *Sort* row in the design grid to specify the field by which records should be sorted. By default, each check box in the *Show* row in the query design grid contains a check mark, meaning the column will be displayed in the query results datasheet. Clear the check mark from a field's *Show* row to hide the column in the query results datasheet.

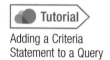

Worldwide Enterprises

What You Will Do Rhonda Trask has requested a list of employees who receive four weeks of vacation. Since you already have the employee names and vacation fields set up in an existing query, you decide to modify the query by adding the vacation criteria and then save the query using a new name.

Tutorial

Adding a Criteria Statement to a Query

Tutorial

Sorting Data and Hiding Fields in Query Results

1. With **3-WEEmployees** open, right-click *BenefitPlans* in the Queries group in the Navigation pane and then click *Design View* at the shortcut menu.

2. Click the File tab, click the *Save As* option, and then click the *Save Object As* option in the *File Types* section of the Save As backstage area.

3. Click the Save As button .

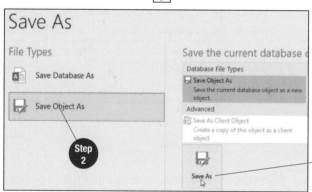

4. Type 4WksVac in the *Save 'BenefitPlans' to* text box at the Save As dialog box and then click OK.

5. With the Query Tools Design tab active, click in the field in the *Criteria* row in the *Vacation* column in the query design grid (the blank field below the check box).

   Before you type a criteria statement, make sure you have placed the insertion point in the *Criteria* row for the field by which you will be selecting records.

6. Type 4 weeks and then press the Enter key.

   The insertion point moves to the field in the *Criteria* row in the next column and Access inserts quotation marks around *4 weeks* in the field in the *Vacation* column. Since quotation marks are required in criteria statements for text fields, Access automatically inserts them if they are not typed into the field in the *Criteria* row.

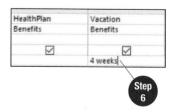

In Brief

Add Criteria Statement to Query
1. Open query in Design view.
2. Click in *Criteria* row in column in which you want to write criterion statement.
3. Type criterion statement.
4. Save revised query.
5. Run query.

Sort Query Results
1. Open query in Design view.
2. Click in *Sort* row in field by which to sort.
3. Click arrow.
4. Click *Ascending* or *Descending*.
5. Save query.
6. Run query.

Hide Column in Query Results
1. Open query in Design view.
2. Click check box in *Show* row in field to be hidden.
3. Save query.
4. Run query.

7 Click in the field in the *Sort* row in the *LastName* column, click the down arrow, and then click *Ascending* at the drop-down list.

8 Click the Run button in the Results group on the Query Tools Design tab.

9 View the query results in the datasheet (the results are sorted in alphabetic order by last name) and then click the View button in the Views group on the Home tab to switch to Design view. (Do *not* click the View button arrow.)

Since Rhonda Trask is interested only in the employee names and vacation, you will instruct Access not to display the other fields in the query results datasheet.

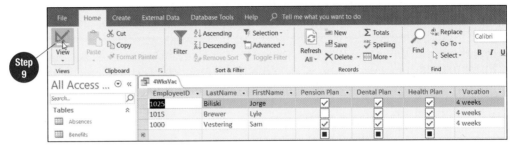

10 Click the check box in the *Show* row in the *PensionPlan* column to remove the check mark.

Removing the check mark instructs Access to hide the column in the query results datasheet.

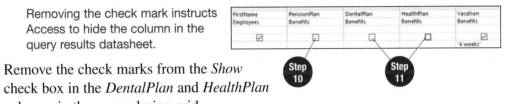

11 Remove the check marks from the *Show* check box in the *DentalPlan* and *HealthPlan* columns in the query design grid.

12 Click the View button to switch to Datasheet view.

The columns for which you removed the check mark from the *Show* check box do not display in the query results. Notice that you displayed the query results datasheet by switching views. Clicking the Run button or switching views achieves the same result.

13 Print the query results datasheet.

14 Close the 4WksVac query. Click the Yes button to save changes to the design of the query.

Check Your Work Compare your work to the model answer available in the online course.

In Addition

Learning More about Criteria Statement

The following are examples of criteria statements for text, number, and date fields showing the proper syntax required by Access. Access inserts the quotation marks (") automatically for text fields and the pound symbols (#) automatically for date fields when a valid entry is typed in a field in the *Criteria* row.

Criterion Statement	Records That Would Be Extracted
"Finance Department"	those with Finance Department in the field
Not "Finance Department"	all except those with Finance Department in the field
"Fan*"	those that begin with Fan and end with any other characters in the field
>15000	those with a value greater than 15,000 in the field
#5/1/21#	those that contain the date May 1, 2021 in the field
>#5/1/21#	those that contain dates after May 1, 2021 in the field

A query may need more than one criterion to select specific records. For example, designing a query that displays records of employees who have enrolled in multiple benefit plans would require more than one criterion. In the query design grid, more than one column can have an entry in the *Criteria* row. Multiple criteria all entered in the same *Criteria* row become an *And* statement wherein each criterion must be met for the record to be selected. For example, the word *Yes* in the *PensionPlan* column and the *DentalPlan* column in the *Criteria* row would mean a record would need to have a check mark in both check boxes in order for Access to display the record in the query results datasheet.

Worldwide Enterprises

What You Will Do   Rhonda Trask is reviewing salaries and has requested a list of employees who work in the North American Distribution Department and earn over $45,000. You will create a new query in Design view to produce the list.

Tutorial

Designing a Query with an *And* Criteria Statement

1. With **3-WEEmployees** open, click the Create tab and then click the Query Design button.

2. At the Show Table dialog box, double-click *Employees* and then click the Close button.

3. Double-click the following fields to add them to the query design grid. *Note: You may have to scroll down the table field list box to see all of the fields*.
    FirstName
    LastName
    Department
    HireDate
    AnnualSalary

4. Click in the field in the *Criteria* row in the *Department* column in the query design grid, type North American Distribution, and then press the Enter key.

5. Position the mouse pointer on the right column boundary line for the *Department* field in the gray header row at the top of the query design grid until the pointer changes to a left-and-right pointing arrow with a vertical line in the middle and then double-click the left mouse button to best fit the column width.

Step 5

Field:	FirstName	LastName	Department	HireDate	AnnualSalary
Table:	Employees	Employees	Employees	Employees	Employees
Sort:					
Show:	☑	☑	☑	☑	☑
Criteria:			"North American Distr		
or:					

Step 4

In Brief

Designing a Query with an *And* Statement
1. Start new query in Design view.
2. Add table(s) and field(s) to query design grid.
3. Click in *Criteria* row in column in which you want to write criterion statement.
4. Type criterion statement.
5. Repeat Steps 3–4 for the remaining criterion fields.
6. Save query.
7. Run query.

6 Click in the field in the *Criteria* row in the *AnnualSalary* column, type >45000, and then press the Enter key.

Placing multiple criterion statements on the same row in the query design grid means that each criterion must be satisfied in order for Access to select the record.

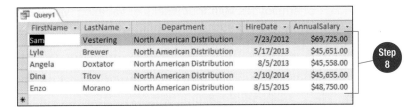

Step 6

Department	HireDate	AnnualSalary
Employees	Employees	Employees
☑	☑	☑
"North American Distribution"		>45000

7 Click the Run button.

8 Review the records selected in the query results datasheet.

The field value for each record in the *Department* column is *North American Distribution* and the field values in the *AnnualSalary* column are all greater than $45,000.

Query1

FirstName	LastName	Department	HireDate	AnnualSalary
Sam	Vestering	North American Distribution	7/23/2012	$69,725.00
Lyle	Brewer	North American Distribution	5/17/2013	$45,651.00
Angela	Doxtator	North American Distribution	8/5/2013	$45,558.00
Dina	Titov	North American Distribution	2/10/2014	$45,655.00
Enzo	Morano	North American Distribution	8/15/2015	$48,750.00

Step 8

9 Click the Save button, type NAHighSalary in the *Query Name* text box, and then press the Enter key or click OK.

10 Print the query results datasheet.

11 Close the NAHighSalary query.

Check Your Work Compare your work to the model answer available in the online course.

In Addition

Learning More about *And* Criteria Statements

The following are additional examples of *And* criteria statements.

Criterion Statement in *PensionPlan* Column	Criterion Statement in *HireDate* Column	Criterion Statement in *AnnualSalary* Column	Records That Would Be Extracted
	>#1/1/2021#	>40000 And <50000	employees hired after January 1, 2021 who earn between $40,000 and $50,000
Yes	>#1/1/2021#	>45000	employees hired after January 1, 2021 who are enrolled in the pension plan and earn over $45,000
No	Between #1/1/2021# And #12/31/2021#	<50000	employees hired between January 1, 2021 and December 31, 2021 who are not enrolled in the pension plan and earn less than $50,000

Multiple criterion statements on different rows in the query design grid become an *Or* statement in which any of the criteria can be met in order for Access to select the record. For example, in this activity, a list of employees who are entitled to either three or four weeks of vacation will be generated. Creating select queries with *Or* statements is often used to generate mailing lists. For example, if a business wants to create mailing labels for customers who live in either Texas or Nevada, the query to select the records needs to be an *Or* statement since the *State* field in a customer table would have a value of *Texas* or *Nevada*. (Having both state names in the same record would not be possible.)

Worldwide Enterprises

What You Will Do   Rhonda Trask has requested a list of employees who receive either three or four weeks of vacation. Since you already have a query created that selected the records of employees with four weeks of vacation, you decide to modify the existing query by adding the second vacation criteria.

Tutorial

Designing a Query with an *Or* Criteria Statement

1. With **3-WEEmployees** open, right-click *4WksVac* in the Queries group in the Navigation pane and then click *Design View* at the shortcut menu.

2. Click the File tab, click the *Save As* option, and then click the *Save Object As* option in the *File Types* section of the Save As backstage area.

3. Click the Save As button.

4. Type 3or4WksVac in the *Save '4WksVac' to* text box at the Save As dialog box and then click OK.

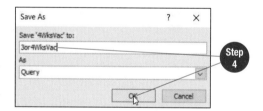

5. With the Query Tools Design tab active, click in the field in the *or* row in the *Vacation* column in the query design grid (the blank row below *"4 weeks"*), type 3 weeks, and then press the Enter key.

   Including a second criterion below the first one instructs Access to display records that meet either of the two criteria.

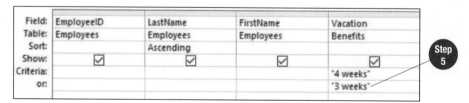

Field:	EmployeeID	LastName	FirstName	Vacation	
Table:	Employees	Employees	Employees	Benefits	
Sort:		Ascending			Step 5
Show:	☑	☑	☑	☑	
Criteria:				"4 weeks"	
or:				"3 weeks"	

6. Click the Run button.

7 View the query results datasheet.

The records that have been selected contain either *4 weeks* or *3 weeks* in the *Vacation* field column.

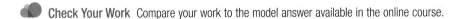

3or4WksVac

EmployeeID ▾	LastName ▾	FirstName ▾	Vacation ▾
1025	Biliski	Jorge	4 weeks
1015	Brewer	Lyle	4 weeks
1023	Bulinkski	Aleksy	3 weeks
1013	Chippewa	Gregg	3 weeks
1005	Deptulski	Roman	3 weeks
1020	Doxtator	Angela	3 weeks
1040	Lafreniere	Guy	3 weeks
1060	McKnight	Donald	3 weeks
1043	Morano	Enzo	3 weeks
1053	O'Connor	Shauna	3 weeks
1003	Ruiz	Celesta	3 weeks
1033	Titov	Dina	3 weeks
1000	Vestering	Sam	4 weeks
*			

Step 7

8 Print the query results datasheet.

9 Click the Save button to save the revised query.

10 Close the 3or4WksVac query.

Check Your Work Compare your work to the model answer available in the online course.

In Addition

Combining *And* and *Or* Criteria Statements

Assume that Rhonda Trask wants to further explore the vacation entitlements for the North American Distribution employees only. Rhonda wants a list of employees who work in the North American Distribution Department *and* have four weeks of vacation *or* who work in the North American Distribution Department *and* have three weeks of vacation. To perform this query, two rows in the query design grid would be used to enter the criteria as shown below. Note that the *Department* column has been added to the query design grid.

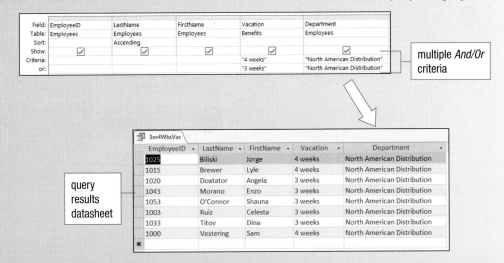

Field:	EmployeeID	LastName	FirstName	Vacation	Department
Table:	Employees	Employees	Employees	Benefits	Employees
Sort:		Ascending			
Show:	☑	☑	☑	☑	☑
Criteria:				"4 weeks"	"North American Distribution"
or:				"3 weeks"	"North American Distribution"

multiple *And/Or* criteria

3or4WksVac

EmployeeID ▾	LastName ▾	FirstName ▾	Vacation ▾	Department ▾
1025	Biliski	Jorge	4 weeks	North American Distribution
1015	Brewer	Lyle	4 weeks	North American Distribution
1020	Doxtator	Angela	3 weeks	North American Distribution
1043	Morano	Enzo	3 weeks	North American Distribution
1053	O'Connor	Shauna	3 weeks	North American Distribution
1003	Ruiz	Celesta	3 weeks	North American Distribution
1033	Titov	Dina	3 weeks	North American Distribution
1000	Vestering	Sam	4 weeks	North American Distribution
*				

query results datasheet

Calculations involving mathematical operations such as adding or multiplying a field value can be included in a query. In a blank field in the *Field* row in Query Design view, type the text that will appear as the column heading, followed by a colon (:) and then the mathematical expression for the calculated values. Field names in the mathematical expression are encased in square brackets. For example, the entry *TotalSalary:[BaseSalary]+[Commission]* would add the value in the field named *BaseSalary* to the value in the field named *Commission*. The result of the equation would be placed in a new column in the query datasheet with the column heading *TotalSalary*. Calculated columns do not exist in the associated table; the values are calculated dynamically each time the query is run. Numeric format and the number of digits after the decimal point for calculated columns are set using the *Format* property box in the Property Sheet task pane in Design view.

What You Will Do Worldwide Enterprises contributes 3% of each employee's annual salary to a registered pension plan. You will create a new query to calculate the employer's annual pension contributions.

Performing
Calculations
in a Query

1 With **3-WEEmployees** open, click the Create tab and then click the Query Design button.

2 At the Show Table dialog box, double-click *Employees* and then click the Close button.

3 Double-click the following fields in the order shown to add them to the query design grid:

    EmployeeID
    FirstName
    LastName
    AnnualSalary

4 Click in the field in the *Field* row to the right of the *AnnualSalary* column in the query design grid.

5 Type PensionContribution:[AnnualSalary]*.03 and then press the Enter key.

6 Position the mouse pointer on the right column boundary line for the *PensionContribution* field in the gray header row at the top of the query design grid until the pointer changes to a left-and-right-pointing arrow with a vertical line in the middle and then double-click the left mouse button to best fit the column width.

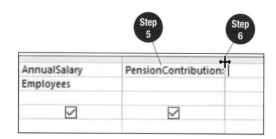

7 Click the Save button. At the Save As dialog box, type PensionCont in the *Query Name* text box and then press the Enter key or click OK.

8 Click the Run button.

9 In the query results datasheet, adjust the column width to best fit the entries in the *PensionContribution* column.

> The values in the calculated column need to be formatted to display a consistent number of decimal places.

10 Switch to Design view.

11 Click in the first field in the *PensionContribution* column in the query design grid.

12 Click the Property Sheet button in the Show/Hide group on the Query Tools Design tab.

> Available properties for the active field display in the Property Sheet task pane at the right side of the work area.

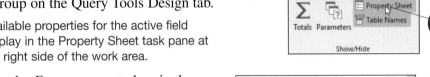

13 Click in the *Format* property box in the Property Sheet task pane, click the down arrow that appears, and then click *Currency* at the drop-down list.

14 Click the Close button in the top right corner of the Property Sheet task pane.

15 Click the Save button and then click the Run button.

16 Click the Totals button in the Records group to add a total row to the bottom of the datasheet and then add a Sum function to the bottom of the *AnnualSalary* and *PensionContribution* field columns. ***Note: Refer to Activity 2.7 if you need assistance with this step.***

17 Print the query results datasheet.

18 Close the PensionCont query. Click Yes when prompted to save changes to the query.

EmployeeID	FirstName	LastName	AnnualSalary	PensionContribution
1000	Sam	Vestering	$69,725.00	$2,091.75
1003	Celesta	Ruiz	$41,875.00	$1,256.25
1005	Roman	Deptulski	$72,750.00	$2,182.50
1013	Gregg	Chippewa	$48,650.00	$1,459.50
1015	Lyle	Brewer	$45,651.00	$1,369.53
1020	Angela	Doxtator	$45,558.00	$1,366.74
1023	Aleksy	Bulinkski	$51,450.00	$1,543.50
1025	Jorge	Biliski	$44,892.00	$1,346.76
1030	Thom	Hicks	$42,824.00	$1,284.72
1033	Dina	Titov	$45,655.00	$1,369.65
1040	Guy	Lafreniere	$45,395.00	$1,361.85
1043	Enzo	Morano	$48,750.00	$1,462.50
1045	Terry	Yiu	$42,238.00	$1,267.14
1050	Carl	Zakowski	$44,387.00	$1,331.61
1053	Shauna	O'Connor	$43,695.00	$1,310.85
1060	Donald	McKnight	$42,126.00	$1,263.78
1063	Charlotte	McPhee	$43,695.00	$1,310.85
1065	Norm	Liszniewski	$43,695.00	$1,310.85
1073	Marsha	Judd	$44,771.00	$1,343.13
1080	Leo	Couture	$43,659.00	$1,309.77
1083	Drew	Arnold	$43,659.00	$1,309.77
1085	Yousef	Armine	$42,177.00	$1,265.31
1090	Maria	Quinte	$42,177.00	$1,265.31
1093	Luis	Vaquez	$42,177.00	$1,265.31
1095	Patrick	Kilarney	$42,177.00	$1,265.31
Total			$1,163,808.00	$34,914.24

Check Your Work Compare your work to the model answer available in the online course.

Activity 3.8 Creating and Formatting a Form

In Access, forms provide a user-friendly interface for viewing, adding, editing, and deleting records. The Form button creates a new form with one mouse click. All fields in the selected table are added to the form in a columnar layout. The form is comprised of a series of objects referred to as *controls*. Each field from the table has a label control and a text box control object. The label control object contains the field name or caption and the text box control object is the field placeholder where data is entered or edited. A form may also contain other controls such as a title, logo, and date and time control object. The form appears in the work area in Layout view, which is used to make changes to the form. Three tabs become active when a form has been created. Use buttons on the Form Layout Tools Design tab to change the theme or add new control objects to the form. The Form Layout Tools Arrange tab contains buttons to rearrange the fields from columnar to tabular or stacked or to otherwise modify the position of the control objects on the form. To make changes to the form's font, font attributes, or other format characteristics, use the Form Layout Tools Format tab. In the Form Wizard, the user is guided through a series of dialog boxes to generate the form, including selecting the fields to be included and the form layout.

Worldwide Enterprises

What You Will Do   You decide to create two forms for the assistant who works with you since she prefers to see only one record at a time while entering data. One form will be used to record employee absences and the other will be used to enter new employee records.

Tutorial
Creating a Form Using the Form Button

Tutorial
Formatting a Form

Tutorial
Creating a Form Using the Form Wizard

1 With **3-WEEmployees** open, click *Absences* in the Tables group in the Navigation pane to select but not open the table.

> In order to create a form using the Form tool, you first select the table or query object upon which to base the new form.

2 Click the Create tab and then click the Form button 📋 in the Forms group.

> Access creates the form using all fields in the table in a vertical layout and displays the form in Layout view with the Form Layout Tools Design tab active.

3 Click the View button 📋 in the Views group on the Form Layout Tools Design tab to switch to Form view.

> When working in a form, you can use the View button to switch back and forth between Form view, where you view the data, and Layout view, where you make changes to the form's appearance and structure.

4 Click the Next record button on the Record Navigation bar a few times to scroll through a few records in Form view.

5 Click the View button 📋 in the Views group on the Home tab to return to Layout view.

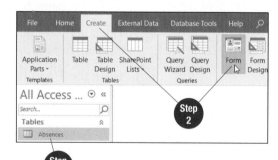

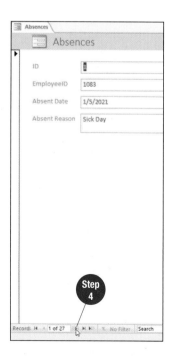

6 Click the Themes button 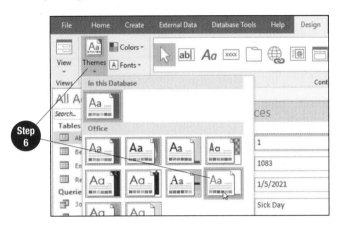 in the Themes group on the Form Layout Tools Design tab and then click *Retrospect* at the drop-down gallery.

> Themes in the Microsoft Office suite are standardized across the applications. A business can apply a consistent look to documents, workbooks, presentations, and databases by using the same theme in each application.

7 Click the Colors button in the Themes group and then click *Green* at the drop-down list.

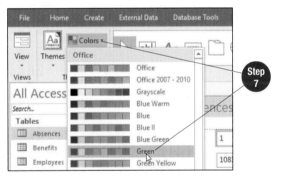

8 Click the Fonts button and then click *Candara* at the drop-down list.

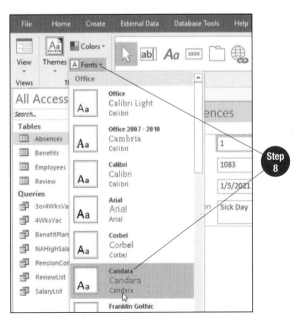

9 Click the Title button 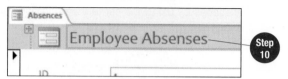 in the Header/Footer group.

Clicking the Title button selects the text *Absences* in the title control object.

10 Type Employee Absences and then press the Enter key.

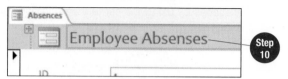

Step 10

11 Click the Form Layout Tools Format tab.

12 Click the *Font Size* option box arrow in the Font group on the Form Layout Tools Format tab and then click *24* at the drop-down list.

13 Press Ctrl + A to select all of the control objects in the form.

14 Click the Font Color button arrow and then click the *Dark Blue* color option (ninth column, bottom row in the *Standard Colors* section).

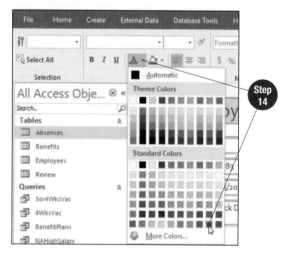

Step 14

15 Close the Absences form. Click the Yes button to save the changes to the form's design and then click OK at the Save As dialog box to accept the default form name *Absences*.

16 Click the Create tab and then click the Form Wizard button 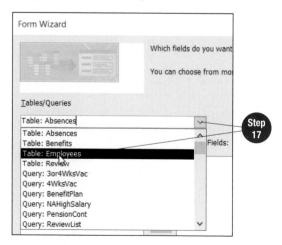 in the Forms group.

17 At the first Form Wizard dialog box, click the *Tables/Queries* option box arrow and then click *Table: Employees* at the drop-down list.

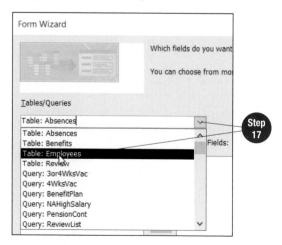

Step 17

In Brief

Create Form Using Form Button
1. Click table name in Navigation pane.
2. Click Create tab.
3. Click Form button.

Create Form Using Form Wizard
1. Click Create tab.
2. Click Form Wizard button.
3. Choose table or query from which to create form.
4. Select fields to include in form.
5. Click Next.
6. Choose form layout.
7. Click Next.
8. Type form title.
9. Click Finish.

18 Click the All Fields button >> to move all of the fields in the *Available Fields* list box to the *Selected Fields* list box and then click the Next button.

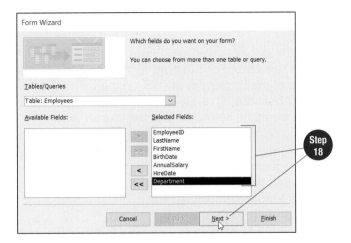

19 At the second Form Wizard dialog box with *Columnar* already selected as the form layout, click the Next button.

20 Click the Finish button at the last Form Wizard dialog box to accept the default title of *Employees* and the option *Open the form to view or enter information*.

21 Click the Next record button on the Record Navigation bar a few times to scroll through a few records in Form view and then close the Employees form.

Check Your Work Compare your work to the model answer available in the online course.

In Addition

Applying Conditional Formatting

Use the Conditional Formatting button to apply formatting to data that meets a specific criterion. For example, conditional formatting can be applied to sales amounts that displays amounts greater than a specified number in a different color. To apply conditional formatting, click the Conditional Formatting button in the Control Formatting group on the Form Layout Tools Format tab and the Conditional Formatting Rules Manager dialog box displays. At this dialog box, click the New Rule button and the New Formatting Rule dialog box displays. Use options in this dialog box to specify the conditional formatting that is to be applied to data in a field that matches a specific condition.

When two tables have been joined by a relationship, fields from a related table can be added to a form. For example, in the Absences form created in the previous activity, the names of the employees are not on the form because the first and last name fields are not in the Absences table. Being able to see the employee's name while entering an absence report would be helpful so that the correct employee ID can be verified as a new record is being added. Add fields from another table to a form using the Field List task pane in Layout view.

Worldwide Enterprises

What You Will Do   You will edit the Absences form to add the employee's first and last names from the Employees table.

Tutorial

Adding an Existing Field to a Form

1 With **3-WEEmployees** open, double-click *Absences* in the Forms group in the Navigation pane to open the form.

2 Click the View button in the Views group on the Home tab to switch to Layout view.

3 With the Form Layout Tools Design tab active, click the Add Existing Fields button ⊞ in the Tools group.

Step 3

The Field List task pane opens at the right side of the work area in one of two states: with one section titled *Fields available for this view* and with a Show all tables hyperlink at the top of the Field List task pane; or, with three sections titled *Fields available for this view*, *Fields available in related tables*, and *Fields available in other tables*. A Show only fields in the current record source hyperlink is at the top of the Field List task pane when it displays three sections.

4 Click Show all tables at the top of the Field List task pane. ***Note: Skip this step if the hyperlink at the top of the Field List task pane on your computer reads Show only fields in the current record source.***

Field List
Step 4
Show all tables
Fields available for this view:
ID
EmployeeID

Clicking Show all tables in the Field List task pane displays two new sections: *Fields available in related tables* and *Fields available in other tables*. Within each section box, each table name displays with an expand button (plus symbol) you can use to display the fields in the related table.

5 Click the expand button next to the Employees table name in the *Fields available in related tables* list box to expand the list.

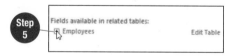
Step 5
Fields available in related tables:
Employees   Edit Table

6 Click to select *FirstName* in the expanded Employees table field list.

7 Position the mouse pointer on the selected *FirstName* field in the Field List task pane, click and hold down the left mouse button, drag the field name to the form between the *EmployeeID* and *Absent Date* fields, and then release the mouse button.

As you drag the field over the form, a pink indicator bar identifies where the field will be positioned between existing fields. When you release the mouse, the field is added to the form. The Field List task pane updates to move the Employees table to the *Fields available for this view* section. By adding a field from the related table, the Employees table is now associated with the Absences table.

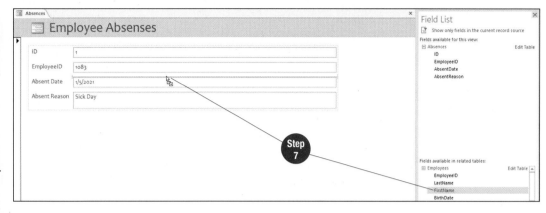

In Brief

**Add Fields to Form
from Another Table**
1. Open Form.
2. Switch to Layout view.
3. Click Add Existing
   Fields button.
4. Click <u>Show all tables</u>.
5. Click expand button
   next to appropriate
   table.
6. Drag field from Field
   List task pane to form.
7. Repeat Step 6 for all
   fields to be added.
8. Close Field List task
   pane.
9. Save and close form.

8   Position the mouse pointer on the *LastName* field in the *Fields available for this
    view* section of the Field List task pane, click and hold down the left mouse button,
    drag the field name to the form between the *FirstName* and *Absent Date* fields, and
    then release the mouse button.

9   Click the Close button in the upper right corner of the Field List task pane.

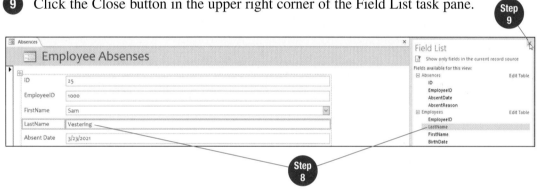

10   Click the Save button to save the revised form design.

11   Switch to Form view and scroll through a few records.

12   Click the New button 🔳 in the Records group on the Home tab, press the Tab key
     to move past the *ID* AutoNumber data type field, type 1063 in the *EmployeeID* field,
     and then press the Tab key or the Enter key.

> Access automatically displays *Charlotte* in the *FirstName* field, *McPhee* in the *LastName*
> field, and *Sick Day* in the *Absent Reason* field. Since the Employees and Absences
> tables are related, Access matches the *EmployeeID* field values and displays the three
> field entries from the primary table.

13   Press the Tab key two times to
     move to the *Absent Date* field, type
     3/30/2021, press the Tab key (this
     selects the current text), type Personal
     leave day in the *Absent Reason* field,
     and then press the Enter key.

14   Close the Absences form.

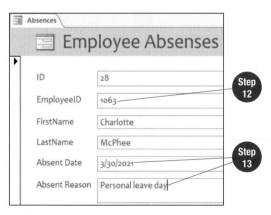

Check Your Work Compare your work to the model answer available in the online course.

A form is comprised of a series of objects referred to as *controls*. Each field from the table has a label control and a text box control object placed side by side, with the label control object placed first. The label control object contains the field name or the caption property text if a caption has been added to the field's properties. The text box control object is the field placeholder where data is entered or edited. The controls can be moved, sized, formatted, or deleted from the form. Use buttons on the Form Layout Tools Design tab and Form Layout Tools Arrange tab to manage control objects. The Property Sheet task pane contains options for changing the width and height of control objects in the form.

Worldwide Enterprises

What You Will Do   You decide to further customize the Absences form by inserting the Worldwide Enterprises logo, resizing control objects, changing the width and height of control objects, and moving control objects.

Tutorial

Managing Control
Objects in a Form

1 With **3-WEEmployees** open, right-click *Absences* in the Forms group in the Navigation pane and then click *Layout View* at the shortcut menu.

2 Click the Logo button in the Header/Footer group on the Form Layout Tools Design tab.

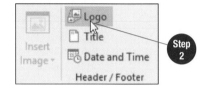
Step 2

3 At the Insert Picture dialog box, navigate to your AccessS3 folder (if AccessS3 is not the current folder) and then double-click the **WELogo-Small** image file.

4 Position the mouse pointer on the right edge of the selected logo control object until the pointer changes to a left-and-right pointing arrow, click and hold down the left mouse button, drag the control to the approximate

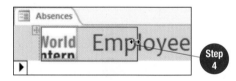
Step 4

width shown at the right, and then release the mouse button. Resize the object as necessary until you can see the entire logo within the control object.

5 Click in the text box control object containing the number *25* (to the right of the *ID* label control object).

6 Click the Property Sheet button in the Tools group.

7 In the Property Sheet task pane with the Format tab active, select the current measurement in the *Width* property box, type 2.2, and then press the Enter key.

8 With the current measurement in the *Height* property box selected, type 0.3 and then press the Enter key.

9 Click in the label control object containing the text *ID*.

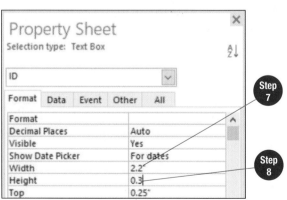
Step 7
Step 8

In Brief

Add Logo to Form
1. Open form in Layout view.
2. Click logo control object.
3. Click Logo button.
4. Navigate to location of graphic file.
5. Double-click graphic file name.

Resize Control Object
1. Open form in Layout view.
2. Select control object.
3. Point to left, right, top, or bottom edge, or to corner.
4. Drag height or width to desired size.

10 In the Property Sheet task pane, select the current measurement in the *Width* property box, type 1.1, and then click the Close button to close the Property Sheet task pane.

11 Click the *Absent Date* label control object, press and hold down the Shift key, click the text box control object containing the date *3/23/2021*, and then release the Shift key.

12 Click the Form Layout Tools Arrange tab.

13 Click the Move Up button in the Move group three times to move the *Absent Date* label control and text box control objects above *EmployeeID*.

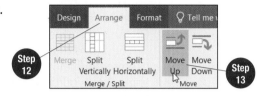

14 Click to select the *EmployeeID* label control object and then click the object again. Move the insertion point between the *e* and the *I* in *EmployeeID*, press the spacebar to insert a space, and then press the Enter key.

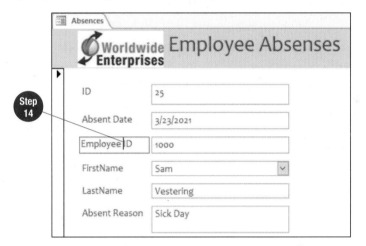

15 Edit the *FirstName* label control to *First Name* and the *LastName* label control to *Last Name* by completing steps similar to those in Step 14.

16 Scroll through the records in the form until you reach the absence report for employee ID 1063 (Charlotte McPhee).

17 Display the form in Print Preview.

18 Click the Columns button in the Page Layout group on the Print Preview tab.

19 Select the current measurement in the *Width* measurement box, type 8, and then click OK.

20 Close Print Preview and then press Ctrl + P to display the Print dialog box.

21 Click *Selected Records(s)* in the *Print Range* section and then click OK.

22 Close the Absences form. Click Yes when prompted to save changes to the form's design.

Check Your Work Compare your work to the model answer available in the online course.

Information from a database can be printed while viewing tables in Datasheet view, while viewing a query results datasheet, or while browsing through forms. In these printouts, all of the fields are printed in a tabular layout for datasheets or in the designated layout for forms. Create a report to specify which fields to print and to have more control over the report layout and format. Access includes a Report button (similar to the Form button) that can be used to generate a report with one mouse click. Reports are generally created for viewing and printing purposes only.

Worldwide Enterprises

What You Will Do Rhonda Trask has requested a hard copy of the NAHighSalary query. You decide to experiment with the Report feature to print the data.

Tutorial
Creating a Report

Tutorial
Formatting a Report

1 With **3-WEEmployees** open, click *NAHighSalary* in the Queries group in the Navigation pane and then click the Create tab.

> Select a table or query before clicking the Report button.

2 Click the Report button in the Reports group.

> Access generates the report using a tabular layout with records displayed in rows. A title, along with the current day, date, and time are placed automatically at the top of the report, as is a container for an image such as a logo at the left of the title text.

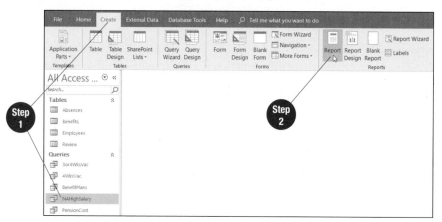

3 Click the Themes button in the Themes group on the Report Layout Tools Design tab and then click the *Integral* theme.

4 Click the Colors button in the Themes group and then click the *Yellow Orange* color option.

5 Click the Title button in the Header/Footer group.

6 Type Employees in North American Distribution, press Shift + Enter to insert a line break, type Earning over $45,000, and then press the Enter key.

7 With the title control object still selected, click the Report Layout Tools Format tab, click the *Font Size* option box arrow in the Font group, and then click *16* at the drop-down list.

In Brief

Create Report Using Report Tool
1. Click object name in Navigation pane.
2. Click Create tab.
3. Click Report button.

8 Click the Report Layout Tools Design tab and then click the Logo button in the Header/Footer group.

9 At the Insert Picture dialog box, navigate to your AccessS3 folder (if AccessS3 is not the current folder) and then double-click **WELogo-Small**. Resize the logo object to the approximate width shown below.

10 Click the Report Layout Tools Format tab and then click the Select All button in the Selection group.

11 Click the Font Color button arrow and then click the *Dark Blue* color option (ninth column, bottom row in the *Standard Colors* section).

12 Click the total amount at the bottom of the *AnnualSalary* field column and then double-click the bottom border to display the entire amount.

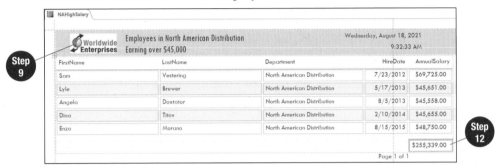

13 Minimize the Navigation pane by clicking the Shutter Bar Open/Close Button at the top of the Navigation pane.

> A dashed line in the middle of the *HireDate* field column indicates a page break. Changing to landscape orientation will allow all of the columns to print on the same page.

14 Click the Report Layout Tools Page Setup tab.

15 Click the Landscape button in the Page Layout group.

> The page break disappears, indicating all columns now fit on one page.

16 Click the Save button. At the Save As dialog box, click OK to accept the default report name *NAHighSalary*.

17 Print and then close the report.

18 Redisplay the Navigation pane by clicking the Shutter Bar Open/Close Button.

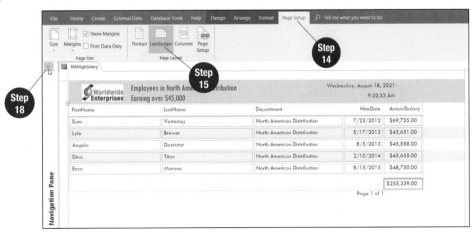

Check Your Work Compare your work to the model answer available in the online course.

Once a report has been created, it can be modified by opening it in Layout view. A report is similar to a form in that it is comprised of a series of controls. Control objects in a report can be resized, moved, and deleted.

Worldwide Enterprises

What You Will Do   Rhonda Trask would like a hard copy of the ReviewList query. You decide to create a new report and move and resize columns to provide a better layout for the report.

Tutorial

Managing Control Objects in a Report

1 With **3-WEEmployees** open, click *ReviewList* in the Queries group in the Navigation pane, click the Create tab, and then click the Report button.

2 Minimize the Navigation pane.

3 With the report displayed in Layout view, click to select the control object containing the *HireDate* column heading. (You may need to scroll to the right to see the column.)

4 Press and hold down the Shift key and then click in any field below the *HireDate* column heading.

> This selects the entire column, as indicated by the gold borders around all of the cells in the column.

5 Position the mouse pointer inside the selected control object containing the *HireDate* column heading until the pointer displays with the four-headed arrow attached.

Steps 3-5

6 Drag the column to the left, between the *LastName* column and the *Supervisor First Name* column.

> A vertical pink indicator bar identifies the location where the column will be placed when you release the mouse.

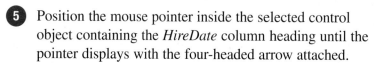

9:41:09 AM			
Supervisor First Name		Supervisor Last Name	HireDate
Sam		Vestering	2/10/2013
Sam		Vestering	5/17/2013
Celesta		Ruiz	8/5/2013
Sam		Vestering	10/10/2013

Step 6

7 Click the *FirstName* column heading to select the control object.

8 Click the Property Sheet button in the Tools group.

9 With the Format tab selected, select the current measurement in the *Width* measurement box, type 0.8, and then click the Close button to close the Property Sheet task pane.

10 Click the *LastName* column heading to select the control object, position the mouse pointer on the right border of the selected control object until the pointer displays as a left-and-right-pointing arrow and then drag to the left until *Lines: 1 Characters: 14* displays at the left side of the Status bar.

In addition to the *Width* measurement box at the Property Sheet task pane, a column width can be changed to a specific width using the line and character numbers that display at the left side of the Status bar.

In Brief

Move Report Columns
1. Open report in Layout view.
2. Click column heading.
3. Press and hold down Shift key and click over data below column heading.
4. Position mouse pointer inside selected column.
5. Click and drag column to new location.
6. Save report.

Resize Report Columns
1. Open report in Layout view.
2. Click column heading.
3. Drag right or left border of selected column heading to change width.
4. Save report.

11 Decrease the width of the *Supervisor First Name* column and the *Supervisor Last Name* column to *Lines 1: Characters 24.*

12 Click the Report Layout Tools Page Setup tab and then change to landscape orientation.

13 Edit the title text to *Employee Review List.*

14 Click to select the logo control object and then press the Delete key.

15 Click to select the control object that contains the current day and date. Position the mouse pointer on the right edge of the selected control until the pointer changes to a left-and-right-pointing arrow. Click and hold down the left mouse button, drag right until the right edge of the date aligns with the right edge of the last column, and then release the mouse button.

Access prints the current day, date, and time on all reports. In businesses with time-sensitive report needs, the inclusion of these controls is important.

16 Click the total amount at the bottom of the *EmployeeID* column, press the Delete key, click the remaining bar, and then press the Delete key again.

17 Click the page number at the bottom of the *Annual Review Date* column and then press the Delete key.

18 Save the report, accepting the default report name of *ReviewList.*

19 Print and then close the ReviewList report.

20 Redisplay the Navigation pane and then close **3-WEEmployees**.

Check Your Work Compare your work to the model answer available in the online course.

In Addition

Understanding Report Sections
A report is divided into five sections, described below.

Report Section	Description
Report Header	Controls in this section are printed once at the beginning of the report, such as the report title.
Page Header	Controls in this section are printed at the top of each page, such as column headings.
Detail	Controls in this section make up the body of the report by printing the data from the associated table or query.
Page Footer	Controls in this section are printed at the bottom of each page, such as the report date and page numbers.
Report Footer	Controls in this section are printed once at the end of the report, such as column totals.

Features Summary

Feature	Ribbon Tab, Group	Button	Keyboard Shortcut
add fields to a form	Form Layout Tools Design, Tools		
create query in Design view	Create, Queries		
Design view	Home, Views		
Form tool	Create, Forms		
Form view	Home, Views		
Form Wizard	Create, Forms		
insert logo in form or report	Form Layout Tools Design, Header/Footer OR Report Layout Tools Design, Header/Footer		
Layout view	Home, Views		
minimize Navigation pane			
Property Sheet task pane	Query Tools Design, Show/Hide OR Form Layout Tools Design, Tools OR Report Layout Tools Design, Tools		Alt + Enter
redisplay Navigation pane			
Report tool	Create, Reports		
run a query	Query Tools Design, Results		
Simple Query Wizard	Create, Queries		

Access

Summarizing Data and Calculating in Forms and Reports

Data Files

Before beginning section work, copy the AccessS4 folder to your storage medium and then make AccessS4 the active folder.

Skills

- Use aggregate functions to calculate statistics in a query
- Summarize data in a crosstab query
- Create a find duplicates query
- Create a find unmatched query
- Insert control objects in a form and report
- Insert a calculation in a form and report
- Sort records in a form
- Group and sort in a report
- Apply conditional formatting to a report
- Create mailing labels
- Compact and repair a database
- Back up a database

Projects Overview

Worldwide Enterprises

Create a query to summarize and group salaries; use a query to find duplicate records and display records from a table that have no matching record in a related table; modify forms and reports to include descriptive text and calculations; and create and modify a report and mailing labels to print names and addresses of distributors.

Performance Threads

Modify a form to include a calculation that shows the rental fee including tax and create a report that prints costume rental revenue grouped by month.

The Waterfront BISTRO

Create a query to summarize and calculate statistics from inventory items and purchases and create a report to summarize catering event revenue.

The online course includes additional training and assessment resources.

Aggregate functions such as Sum, Avg, Min, Max, and Count can be used in a query to calculate statistics. When an aggregate function is used, Access displays one row in the query results datasheet with the result for each statistic in the query. To display the aggregate function list, click the Totals button in the Show/Hide group on the Query Tools Design tab. Access adds a *Total* row to the query design grid with a drop-down list from which a function can be selected. Using the *Group By* option in the *Total* drop-down list, a field can be added to the query upon which Access will group records for statistical calculations.

What You Will Do Rhonda Trask, human resources manager, has asked for statistics on the salaries currently paid to employees. You will create a new query and use aggregate functions to find the total of all salaries, the average salary, and the maximum and minimum salary. In a second query, you will calculate the same statistics by department.

Using Aggregate Functions

1. Open **4-WEEmployees** and enable the content, if necessary.

2. Click the Create tab and then click the Query Design button.

3. At the Show Table dialog box with the Tables tab selected, double-click *Employees* and then click the Close button.

4. Double-click *AnnualSalary* in the Employees table field list box to add the field name to the first field in the *Field* row in the query design grid. Double-click *AnnualSalary* again to add the field name to the second field in the *Field* row. Repeat two more times so that *AnnualSalary* appears in the first four fields in the *Field* row.

   The field upon which the statistics are to be calculated is added to the query design grid once for each aggregate function you want to use.

Step 4

5. Click the Totals button Σ in the Show/Hide group on the Query Tools Design tab.

   A *Total* row is added to the query design grid between the *Table* and *Sort* rows with the default option *Group By*.

6. Click in the first *AnnualSalary* field in the *Total* row in the query design grid, click the arrow and then click *Sum* at the drop-down list.

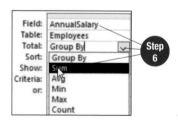

Step 6

7. Click in the second *AnnualSalary* field in the *Total* row, click the arrow, and then click *Avg* at the drop-down list.

8. Change the *Group By* option to *Max* for the third *AnnualSalary* field in the *Total* row.

9. Change the *Group By* option to *Min* for the fourth *AnnualSalary* field in the *Total* row.

10. Click the Save button, type SalaryStats in the *Query Name* text box at the Save As dialog box, and then press the Enter key or click OK.

In Brief

Use Aggregate Functions

1. Create new query in Design view.
2. Add required table(s).
3. Close Show Table dialog box.
4. Add field for each function.
5. Click Totals button.
6. Change *Total* option to specific function in each field column.
7. Save query.
8. Run query.

11 Click the Run button.

Access calculates the Sum, Avg, Max, and Min functions for all salary values and displays one row with the results. Access assigns column headings in the query results datasheet using the function name, the word *Of*, and the field name from which the function has been derived; for example, *SumOfAnnualSalary*.

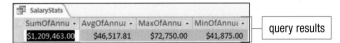

query results

12 Adjust all column widths to best fit the entries in the query results datasheet and then print the datasheet.

13 Switch to Design view.

14 Click the File tab and then click the *Save As* option.

15 Click the *Save Object As* option in the *File Types* section and then click the Save As button.

16 Type SalaryStatsByDept in the *Save 'SalaryStats' to* text box and then press the Enter key or click OK.

17 Click the Query Tools Design tab and then double-click *Department* in the Employees table field list box. ***Note: You may have to scroll down the table field list box to find the* Department *field*.**

The *Department* field is added to the query design grid with *Total* automatically set to *Group By*. Adding this field produces a row in the query results datasheet in which Access calculates the aggregate functions for each department.

18 Run the query, change the orientation to landscape, and then print the query results datasheet.

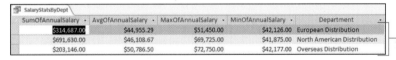

Steps 17-18

19 Close the SalaryStatsByDept query. Click Yes when prompted to save changes to the query design.

Check Your Work Compare your work to the model answer available in the online course.

In Addition

Using the Caption Property Box

Access assigned column headings in the query results datasheet using the function name, the word *Of*, and the field name from which the function was derived. These default column headings can be changed using the *Caption* property box in the Property Sheet task pane. Display the task pane by clicking the Property Sheet button in the Show/Hide group on the Query Tools Design tab. Click in a row in a column in the query design grid, click in the *Caption* property box, and then type a new name for the column heading. After renaming each column heading, close the Property Sheet task pane by clicking the Close button in the upper right corner of the task pane.

A crosstab query calculates aggregate functions such as Sum and Avg in which field values are grouped by two fields. Access includes a wizard with the steps required to create the query. The first field selected causes one row to display in the query results datasheet for each group. The second field selected displays one column in the query results datasheet for each group. The third field specified is the numeric field to be summarized. The cell at the intersection of each row and column holds a value that is the result of the specified function for the designated row and column group. For example, to find the total sales achieved by each salesperson by state, each row in the query results could be used to display a salesperson's name with the state names in columns. Access summarizes the total sales for each person for each state and shows the results in a spreadsheet-type format.

What You Will Do Rhonda Trask wants to find out the salary cost that has been added to the payroll each year by department. You will use a crosstab query to calculate the total value of annual salaries for new hires in each year by each department.

Creating a Crosstab Query

1 With **4-WEEmployees** open, click the Create tab and then click the Query Wizard button.

2 Click *Crosstab Query Wizard* in the New Query dialog box and then click OK.

> The fields that you want to use for grouping must all exist in one table or query. In situations in which the fields that you want to group by are in separate tables, you would first create a new query that contains the fields you need and then start the Crosstab Query Wizard. In this activity, all three fields you need are in one table.

3 At the first Crosstab Query Wizard dialog box with *Tables* selected in the *View* section, click *Table: Employees* in the list box and then click the Next button.

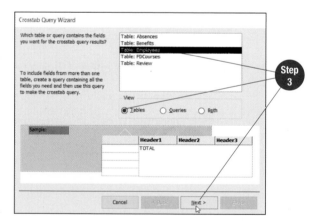

4 At the second Crosstab Query Wizard dialog box, double-click *Department* in the *Available Fields* list box to move the field to the *Selected Fields* list box and then click the Next button.

> At the second Crosstab Query Wizard dialog box, you choose the field in which the field's values become the row headings in the query results datasheet.

In Brief

Create Crosstab Query
1. Click Create tab.
2. Click Query Wizard button.
3. Click *Crosstab Query Wizard* and click OK.
4. Choose table or query name and click Next.
5. Choose field for row headings and click Next.
6. Choose field for column headings and click Next.
7. Choose numeric field to summarize and function to calculate and click Next.
8. Type query name and click Finish.

5 At the third Crosstab Query Wizard dialog box, click *HireDate* in the field list box and then click the Next button.

Whenever a date/time field is chosen for the column headings, Access displays a dialog box asking you to choose the time interval to summarize by, with the default option set to *quarter*.

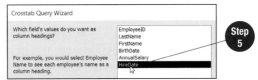

6 At the fourth Crosstab Query Wizard dialog box, click *Year* in the list box and then click the Next button.

At the fourth dialog box, you choose the field in which the field's values become the column headings in the query results datasheet.

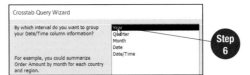

7 At the fifth Crosstab Query Wizard dialog box, click *AnnualSalary* in the *Fields* list box and then click *Sum* in the *Functions* list box.

At the fifth Crosstab Query Wizard dialog box, you select the numeric field to be summarized and the function to be used to calculate the values.

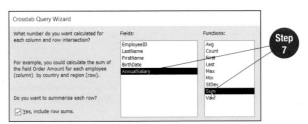

8 Notice how the datasheet layout displays in the *Sample* section and then click the Next button.

9 Select the existing text in the *What do you want to name your query?* text box, type PayrollByDeptByYear, and then click the Finish button.

10 Minimize the Navigation pane and adjust each column's width to best fit the entries.

The query results datasheet displays as shown below. Notice the total column next to each department name and the columns that break down the totals to show the payroll cost for each year.

Department	Total Of AnnualSalary	2012	2013	2014	2015	2016	2017	2018	2019	2020	2021
European Distribution	$314,687.00		$100,100.00		$42,238.00	$44,387.00	$42,126.00			$43,659.00	$42,177.00
North American Distribution	$691,630.00	$111,600.00	$136,101.00	$91,310.00	$48,750.00		$43,695.00	$87,390.00	$44,771.00	$43,659.00	$84,354.00
Overseas Distribution	$203,146.00	$72,750.00		$42,824.00	$45,395.00						$42,177.00

11 Print the query results datasheet in landscape orientation.

12 Close the PayrollByDeptByYear query. Click Yes to save changes to the layout.

13 Redisplay the Navigation pane.

 Check Your Work Compare your work to the model answer available in the online course.

A find duplicates query searches a table or query for duplicate values within a designated field or group of fields. Create this type of query if a record, such as a product record, may have been entered two times, such as under two different product numbers. Other applications for this type of query are included in the In Addition section at the end of this activity. Access provides the Find Duplicates Query Wizard to help build the query by making selections in a series of dialog boxes.

Worldwide Enterprises

What You Will Do   You printed new ID cards for employees and received one more card than the number of employees. You suspect that a duplicate ID card was printed. You will use a find duplicates query to check for a duplicated record in the Employees table.

Tutorial

Creating a Find
Duplicates Query

1 With **4-WEEmployees** open, click the Create tab and then click the Query Wizard button.

2 Click *Find Duplicates Query Wizard* in the New Query dialog box and then click OK.

3 With *Tables* selected in the *View* section of the first Find Duplicates Query Wizard dialog box, click *Table: Employees* in the list box and then click the Next button.

> At the first Find Duplicates Query Wizard dialog box, you choose the table or query in which you want Access to look for duplicate field values.

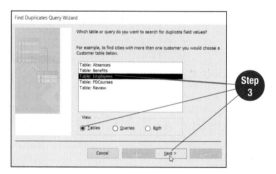

4 Double-click *LastName* and *FirstName* in the *Available fields* list box to move them to the *Duplicate-value fields* list box and then click the Next button.

> In the second Find Duplicates Query Wizard dialog box, you choose the fields that may contain duplicate field values. Since *EmployeeID* is the primary key field in the Employees table, you know that it is not possible for an employee record to be duplicated using the same employee number; therefore, you will use the name fields to check for duplicates.

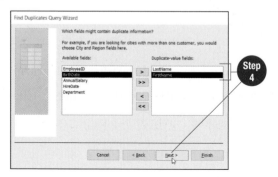

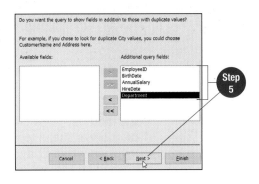

Step 5

5 Move all of the fields from the *Available fields* list box to the *Additional query fields* list box and then click the Next button.

> If an employee record has been duplicated, you want to see all of the fields to ensure that the information in both records is exactly the same. If not, you need to figure out which record contains the accurate information before deleting the duplicate.

6 With the text already selected in the *What do you want to name your query?* text box, type DupRecordCheck and then click the Finish button.

> The query results datasheet displays, showing that Dina Titov has two records in the Employees table under two different employee numbers.

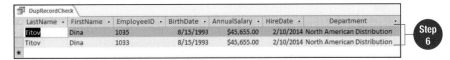

Step 6

7 Print the query results datasheet in landscape orientation.

8 Move the mouse pointer in the record selector bar next to the record with the employee ID 1035 until the pointer changes to a right-pointing black arrow. Right-click the record selector bar and then click *Delete Record* at the shortcut menu.

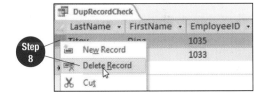

Step 8

9 Click Yes to confirm the record deletion.

10 Close the DupRecordCheck query.

11 Double-click *DupRecordCheck* in the Queries group in the Navigation pane to reopen the query. The query results datasheet is now blank. Since you deleted the duplicate record for Dina Titov in Step 8, duplicate records no longer exist in the Employees table.

12 Close the DupRecordCheck query.

Check Your Work Compare your work to the model answer available in the online course.

In Addition

Understanding When to Use a Find Duplicates Query

In this activity, a find duplicates query was used to locate and then delete an employee record that was entered two times. A find duplicates query has several other applications. Consider the following examples:
- Find the records in an Orders table with the same customer number so that you can identify your most loyal customers.

- Find the records in a Customer table with the same last name and mailing address so that you send only one mailing to a household, saving on printing and postage costs.
- Find the records in an Expenses table with the same employee number so that you can see which employee is submitting the most claims.

Use a find unmatched query to have Access compare two tables and produce a list of the records in one table that have no matching record in the other related table. This type of query is useful to produce lists such as customers who have never placed an order, invoices with no payment, or employees with no absences. Access provides the Find Unmatched Query Wizard with the steps for building the query.

What You Will Do You are not sure that benefits have been entered for all employees. You decide to create a find unmatched query to make sure that a record exists in the Benefits table for all employees.

Creating a Find Unmatched Query

1 With **4-WEEmployees** open, click the Create tab and then click the Query Wizard button.

2 Double-click the *Find Unmatched Query Wizard* option in the New Query dialog box.

3 With *Tables* selected in the *View* section of the first Find Unmatched Query Wizard dialog box, click *Table: Employees* in the list box and then click the Next button.

At the first Find Unmatched Query Wizard dialog box, you choose the table or query in which you want to view records in the query results datasheet. If an employee is missing a record in the Benefits table, you will need the employee's name and employee number, which are in the Employees table.

4 At the second Find Unmatched Query Wizard dialog box, click *Table: Benefits* and then click the Next button.

At the second dialog box, you choose the table or query that you want Access to compare with the first table selected. For Access to compare records, you need to specify the field in each table that would have matching field values.

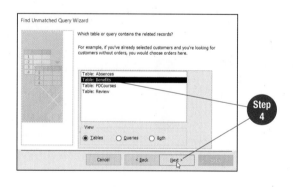

5 With *EmployeeID* already selected in the *Fields in 'Employees'* and *Fields in 'Benefits'* list boxes at the third Find Unmatched Query Wizard dialog box, click the Next button.

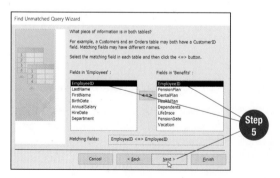

6 At the fourth Find Unmatched Query Wizard dialog box, double-click *EmployeeID*, *LastName*, and *FirstName* to move the fields from the *Available fields* list box to the *Selected fields* list box and then click the Next button.

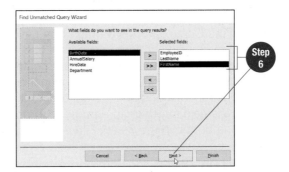

7 With the text already selected in the *What would you like to name your query?* text box, type UnmatchedBenefits and then click the Finish button.

8 Look at the three records displayed in the query results datasheet. These are the employee records for which no record with a matching employee number exists in the Benefits table.

9 Print the query results datasheet and then close the UnmatchedBenefits query.

10 Open the Benefits table and then add the records shown in Figure 4.1 to the table.

11 Close the Benefits table.

12 Double-click *UnmatchedBenefits* in the Queries group in the Navigation pane. The query results datasheet is now blank since all employees now have a matching record in the Benefits table.

13 Close the UnmatchedBenefits query.

Figure 4.1 Step 10

EmployeeID	Pension Plan	Dental Plan	Health Plan	Dependents	Life Insurance	Pension Date	Vacation
1045	Yes	No	Yes	3	$150,000	01-May-15	3 weeks
1080	Yes	No	No	0	$100,000	01-Feb-20	2 weeks
1085	Yes	Yes	Yes	4	$185,000	01-Jan-21	1 week

Check Your Work Compare your work to the model answer available in the online course.

The gallery in the Controls group on the Form Layout Tools Design tab contains buttons for inserting control objects in a form, such as a title, the date and time, and a picture or other image, to name a few. A label control object contains text that is not associated with an existing field in a table or query. Think of a label control object like a text box that can be used to add instructions or other explanatory text for the users of a form. To sort data in a form, click in the field by which to arrange records and then click the Ascending button or the Descending button in the Sort & Filter group on the Home tab.

Worldwide Enterprises

What You Will Do   Worldwide Enterprises reimburses employees for tuition paid for work-related courses, provided the employee achieves a grade of C- or higher. A form has already been created to enter the records into the PDCourses table. You decide to add objects to the form to enhance its appearance and functionality.

Tutorial

Inserting Control Objects

1   With **4-WEEmployees** open, double-click *PDCourses* in the Forms group in the Navigation pane to open the form.

2   Scroll through a few records in the form to become familiar with the form's data and purpose. Display the first record in the form and then click the View button in the Views group on the Home tab to switch to Layout view.

> In the next steps, you will insert a label control object in the form to add an explanation about the minimum grade requirement.

3   Click the Label button Aa in the gallery in the Controls group on the Form Layout Tools Design tab.

Step 3

> When a control button has been selected, the pointer changes to crosshairs with an icon attached that indicates the type of control object to be created. A label control object displays an uppercase *A* as the icon.

4   Position the crosshairs with the label icon attached at the left edge of the form, just below the *Date Reimbursed* field.

Step 4

> A vertical pink indicator bar appears just below the *Date Reimbursed* label to show you where the new control object will be placed.

5   Click to insert a label control object. The insertion point is positioned in the label control object. Type Minimum Grade of C- is Required. and then press the Enter key.

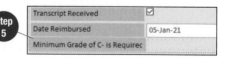

Step 5

> The column width automatically expands to accommodate the length of the text entry. Press Shift + Enter to insert a line break when you want to move down one line in the label control object. A line break can help readability of long entries. In the next steps, you will add a picture to the form to create visual interest.

6   Click the Image button in the gallery in the Controls group.

Step 6

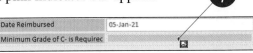
Step
7

7 Position the crosshairs with the image icon attached below the date entry in the *Date Reimbursed* field until a pink indicator bar appears inside a rectangle next to the label control object and then click the left mouse button.

8 At the Insert Picture dialog box, navigate to your AccessS4 folder and then double-click **Classroom**.

9 Position the mouse pointer on the bottom of the image control object until the pointer displays as an up-and-down-pointing arrow and then drag the mouse down approximately 2 inches to resize the image as shown at the right.

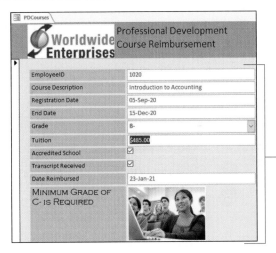

Step
9

10 Click to select the label control object to the left of the picture and then click the Form Layout Tools Format tab.

11 Change the font to Copperplate Gothic Light and the font size to 14 points.

12 Click outside the label control object to deselect the object.

13 Save the form and then switch to Form view.

14 At the first record displayed in Form view, click in the *Tuition* field and then click the Ascending button in the Sort & Filter group on the Home tab.

> The records are rearranged to display from the lowest to highest tuition reimbursed.

first record displayed when form is sorted in ascending order on the *Tuition* field in Step 14

15 Print the first record only.

16 Close the PDCourses form, saving changes if prompted.

Check Your Work Compare your work to the model answer available in the online course.

Create a calculated control object containing a formula by using the Text Box button. Like a calculated field in a query, a calculated control object in a form is not stored as a field—each time the form is opened, the results are calculated dynamically. Display a form in Layout view to add a text box control object.

What You Will Do Worldwide Enterprises pays its employees 4% of their annual salary each year as vacation pay. You decide to show the vacation pay calculation within the Employees form.

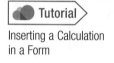

Inserting a Calculation in a Form

1. With **4-WEEmployees** open, right-click *Employees* in the Forms group in the Navigation pane and then click *Layout View* at the shortcut menu.

2. Click the Text Box button abl in the Controls group on the Form Layout Tools Design tab.

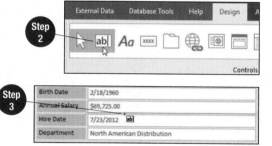

3. Position the crosshairs with the text box icon attached between the *Annual Salary* and *Hire Date* fields until a horizontal pink indicator bar displays between the field boxes and then click the left mouse button.

With the Text Box button selected, clicking in the form inserts two control objects: a label control object and a text box control object.

4. Click to select the empty text box control object (white box below $69,725.00) and then click the Property Sheet button in the Tools group on the Form Layout Tools Design tab.

5. Click the Data tab in the Property Sheet task pane.

6. Click in the *Control Source* property box, type =[AnnualSalary]*.04, and then press the Enter key.

Similar to a query, a formula is typed by placing field names in square brackets. The field name for the salary field is *AnnualSalary* (with no spaces between the words). In the form, recall that the label controls next to the fields can contain spaces to improve readability and appearance; however, in a formula, you must refer to the proper field name in which the data resides in the table. The Property Sheet task pane can be widened to display the entire formula.

7. Click the Close button in the upper right corner of the Property Sheet task pane to close the task pane.

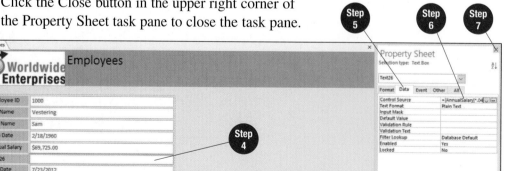

8 With the text box control object still selected (contains *2789*), click the Form Layout Tools Format tab.

9 Click the Apply Currency Format button $ in the Number group.

> Applying currency formatting changes the amount in the text box control object to *$2,789.00*.

10 Click the label control object adjacent to the calculated text box control object (currently displays *Text##* [where *##* is the number of the label object]).

11 Double-click the text in the label control object to select it, type Vacation Pay, and then press the Enter key.

12 Position the mouse pointer on the right edge of the selected *Vacation Pay* label control object until the pointer displays as a left-and-right pointing arrow and then drag right approximately 0.25 inch to widen the label column. (This changes the width of the entire column.)

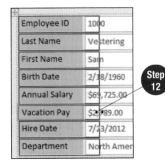

13 Click in a blank area outside the label control object to deselect the object.

14 Switch to Form view and then scroll through the records, viewing the calculated control object for each record.

15 Display the first record in the form and then print the selected record.

16 Save and then close the Employees form.

Check Your Work Compare your work to the model answer available in the online course.

In Addition

Using the Expression Builder Dialog Box

In addition to typing a calculation into the *Control Source* property box, a calculation can be created using the Expression Builder dialog box. Display the dialog box by clicking in the *Control Source* property box in the Property Sheet task pane and then clicking the Build button at the right side of the property box. To insert the formula for the vacation pay, scroll down the *Expression Categories* list box and then double-click *AnnualSalary*. The Expression Builder adds the field name in square brackets to the text box in the top portion of the dialog box. To complete the calculation, type *.04 and then press the Enter key. Close the Expression Builder dialog box by clicking the Close button in the upper right corner.

Labels, images, and calculated control objects can be added to a report using similar techniques as those learned in the previous activities with forms. A calculation can be inserted in a report using the Text Box button in the gallery in the Controls group on the Report Layout Tools Design tab. Insert a calculation in a text box in the same manner as inserting a calculation in a form. Display the Property Sheet task pane and then type the calculation in the *Control Source* property box.

Worldwide Enterprises

What You Will Do   Rhonda Trask estimates that Worldwide Enterprises incurs an additional 22% of an employee's annual salary to pay for the employee's benefit plan. You will create a query and then a report and insert a calculation to show the estimated benefit cost by employee.

Tutorial

Inserting a Calculation in a Report

1 With **4-WEEmployees** open, click the Create tab, click the Query Design button, and then create a new query as follows:
   - Add the Employees table to the query design grid.
   - Add the following fields in this order: *EmployeeID*, *FirstName*, *LastName*, *HireDate*, *Department*, and *AnnualSalary*.
   - Save the query and name it *EmployeeList*.

2 Run the query, view the query results datasheet, and then close the query.

3 Click *EmployeeList* in the Queries group in the Navigation pane, click the Create tab, and then click the Report button in the Reports group.

4 Minimize the Navigation pane.

5 Click the Report Layout Tools Page Setup tab and then click the Landscape button in the Page Layout group.

6 Click to select the control object containing the *FirstName* column heading and then drag the right edge of the *FirstName* control object left until *Lines: 1 Characters: 15* displays at the left side of the Status bar.

7 Click to select the control object containing the *LastName* column heading and then drag the right edge of the *LastName* control object left until *Lines 1: Characters: 15* displays at the left side of the Status bar.

8 Click the Report Layout Tools Design tab.

9 Click the Text Box button in the Controls group and then position the crosshairs with the text box icon attached at the right of the *AnnualSalary* column heading until a vertical pink indicator bar appears.

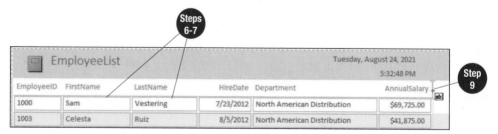

10 Click the left mouse button to insert a new column.

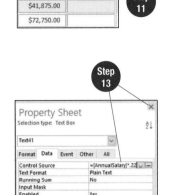

11 Click in the empty cells below the *Text##* field column heading (where ## is the number Access has assigned to the text box control object) to select the cells below the heading.

12 Click the Property Sheet button in the Tools group on the Report Layout Tools Design tab.

13 Click the Data tab, click in the *Control Source* property box, type =[AnnualSalary]*.22, and then close the Property Sheet task pane.

14 Click the Report Layout Tools Format tab and then click the Apply Currency Format button in the Number group.

15 Double-click the label control object at the top of the calculated column, delete the current text, type Benefit Cost, and then press the Enter key.

16 Drag the right edge of the *Benefit Cost* label control object to the right until *Lines: 1 Characters: 14* displays at the left side of the Status bar.

17 Click to select the logo control at the top left of the report and then press the Delete key to delete the control object.

18 Double-click the report title *EmployeeList*, delete the current text, type Employee Benefit Costs, and then press the Enter key.

19 Add a space between the words in the field names for those column headings that do not currently have a space.

20 Scroll to the bottom of the report and then select and delete the total and the line at the bottom of the *Annual Salary* field column. (Select the cell and then press the Delete key. Click the remaining line and then press the Delete key.)

21 Display the report in Print Preview and then return to Layout view. Widen any columns that displayed pound symbols (#) in Print Preview.

22 Click the table select handle above and to the left of the *Employee ID* column heading to select all records in the report.

23 Click the Report Layout Tools Arrange tab, click the Control Padding button in the Position group, and then click *None* at the drop-down list.

> Control padding adds extra space between rows in a report. Removing the padding allows the report to better fit on one page.

24 Click to select the current day and date control object and the time control object and then drag the right edge of the selected controls to the right until the control aligns with the right edge of the *Benefit Cost* calculated column.

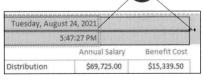

25 Save the report, accepting the default name *EmployeeList*, and then print and close the report.

26 Redisplay the Navigation pane.

Check Your Work Compare your work to the model answer available in the online course.

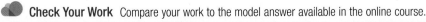

Use the Group, Sort, and Total pane in Layout view to group together records with a similar characteristic. For example, the records in the report can be arranged by department so that all of the employees in the North American Distribution Department are shown together. By default, when a report is grouped, the records are sorted by the grouped field. This sort order can be changed or multiple-level sorting can be specified to sort by another field within a group (such as last name). Display the Group, Sort, and Total pane by clicking the Group & Sort button in the Grouping & Totals group on the Report Layout Tools Design tab.

What You Will Do Rhonda Trask has asked for a new report that groups the salaries by department. You decide to create a report based on the EmployeeList query and add grouping to the report.

Grouping and Sorting
Records in a Report

1 With **4-WEEmployees** open, click *EmployeeList* in the Queries group in the Navigation pane, click the Create tab, and then click the Report button.

2 Minimize the Navigation pane.

3 Click the Report Layout Tools Page Setup tab and then click the Landscape button.

4 Click the Report Layout Tools Design tab and then click the Group & Sort button in the Grouping & Totals group. *Note: Skip this step if the Group, Sort, and Total pane is already open at the bottom of the work area.*

> The Group, Sort, and Total pane opens at the bottom of the work area. The Group & Sort button is a toggle button that opens or closes the pane.

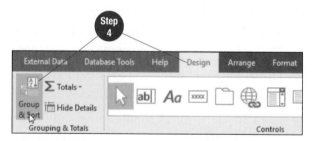

5 Click the Add a group button in the Group, Sort, and Total pane.

> Access adds a grouping level row in the Group, Sort, and Total pane and a list box of available fields displays next to *select field* in which you choose the field by which the records should be grouped.

6 Click *Department* in the list box of available fields in the Group, Sort, and Total pane.

> Access reorganizes the records in the report and displays a group for each department name with the associated employee records for each group.

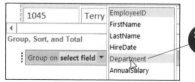

In Brief

Group Data in Report
1. Open report in Layout view.
2. If necessary, click Report Layout Tools Design tab.
3. Click Group & Sort button.
4. Click Add a group button.
5. Click field name by which to group.
6. Close Group, Sort, and Total pane.
7. Save report.

Sort Data in Report
1. Open report in Layout view.
2. If necessary, click Report Layout Tools Design tab.
3. Click Group & Sort button.
4. Click Add a sort button.
5. Click field name by which to sort.
6. Close Group, Sort, and Total pane.
7. Save report.

7 Observe that the department names are automatically sorted in ascending order.

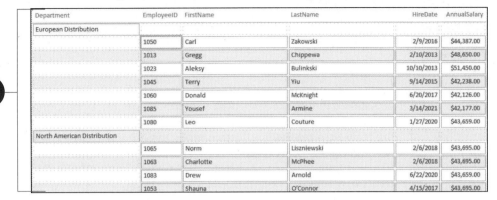

Step 7

8 Click the Add a sort button in the Group, Sort, and Total pane.

Access inserts a sorting row below the grouping level row and a list box of available fields displays from which you choose the field by which to sort the records within each group.

9 Click *LastName* in the list box of available fields.

The records are now arranged in ascending order by last name within each group.

Step 9

10 Click the Group & Sort button in the Grouping & Totals group on the Report Layout Tools Design tab to close the Group, Sort, and Total pane.

11 Make the following changes to the report:
- Apply the Integral theme.
- Delete the logo control object.
- Delete the total and line at the bottom of the *AnnualSalary* field column.
- Change the report title to *Employee Salaries by Department*.
- Add spaces between words in the column headings.
- Adjust column widths to better fit data on the page.
- Change the Control Padding to None.
- Move the date and time controls to align at the right edge of the report.

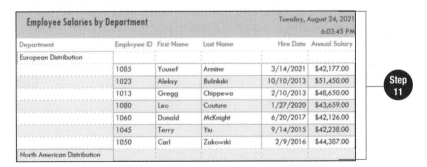

Step 11

12 Save the report and name it *EmployeesByDept*.

13 Print and close the report and then redisplay the Navigation pane.

Check Your Work Compare your work to the model answer available in the online course.

Activity 4.9 Applying Conditional Formatting to a Report

Conditional formatting can be used to apply font formatting options to field values in a report that meet a specific criterion. For example, a different font color can be applied to all salaries that are greater than a certain value. Click the Conditional Formatting button on the Report Layout Tools Format tab to open the Conditional Formatting dialog box, click the New Rule button, and then specify the criterion and the formatting to apply at the New Formatting Rule dialog box.

What You Will Do   Rhonda Trask has requested a report that shows the employees who have participated in the tuition reimbursement program. She is specifically interested in the courses that were reimbursed that cost over $1,000.00.

Applying Conditional Formatting to a Report

1. With **4-WEEmployees** open, click the Create tab, click the Query Design button, and then create a new query as follows:
   - Add the Employees and the PDCourses tables to the query window.
   - Add the following fields in order: *EmployeeID, FirstName, LastName, CourseDescr, Grade, Tuition,* and *Reimbursed.*
   - Save the query and name it *TuitionReimbursed.*

2. Run the query, view the query results datasheet, and then close the query.

3. Create a new report with the TuitionReimbursed query.

4. Minimize the Navigation pane.

5. Change the report to landscape orientation and adjust the column widths so that all of the fields fit on one page.

6. Position the mouse pointer over the values in the *Tuition* field column and then click to select the column.

7. Click the Report Layout Tools Format tab and then click the Conditional Formatting button [icon] in the Control Formatting group.

   The Conditional Formatting Rules Manager dialog box opens, in which you can create a new criterion, referred to as a *rule,* or edit or delete an existing rule.

8. Click the New Rule button in the Conditional Formatting Rules Manager dialog box.

   The New Formatting Rule dialog box opens. By default, the new rule type is *Check values in the current record or use an expression.*

9. Click the second option box arrow (currently displays *between*) in the *Edit the rule description* section and then click *greater than or equal to* at the drop-down list.

10. Click in the blank text box next to *greater than or equal to* and then type 1000.

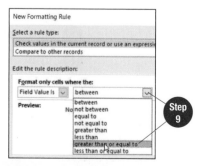

In Brief

Apply Conditional Formatting to a Report
1. Open report in Layout view.
2. Select field values to format.
3. Click Report Layout Tools Format tab.
4. Click Conditional Formatting button.
5. Click New Rule button.
6. Change *between* list option as required.
7. Type criterion text or value.
8. Change formatting options as required.
9. Click OK.
10. Click OK.

11 Click the Font color button arrow below the text box and then click the *Dark Red* color option (first column, last row in the *Standard Colors* section).

12 Click the Bold button.

13 Click OK to close the New Formatting Rule dialog box.

The new rule is shown in the *Rule* list in the Conditional Formatting Rules Manager dialog box.

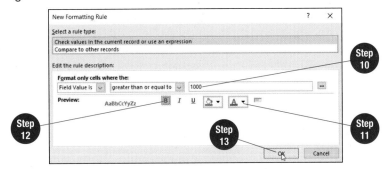

14 Click OK to close the Conditional Formatting Rules Manager dialog box.

The rule is applied to the selected column when you close the dialog box.

15 Notice that the values in the *Tuition* field column that are over $1,000.00 appear in bold, dark red font.

16 Make the following changes to the report:
- Delete the logo control object.
- Change the report title to *Employee Tuition Reimbursement Report*.
- Add spaces between words in the column headings whose field names are compound words.
- Move the date and time controls to align at the right edge of the report.
- Delete the total amount and line at the bottom of the *Tuition* field column.

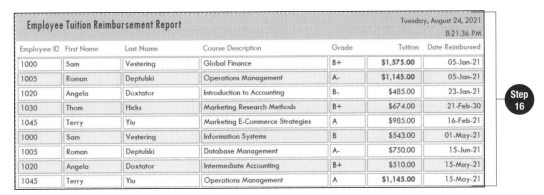

17 Save the report with the default name *TuitionReimbursed*.

18 Print and then close the report.

19 Redisplay the Navigation pane and then close **4-WEEmployees**.

Check Your Work Compare your work to the model answer available in the online course.

Often, an Access database contains the names and addresses of customers, employees, vendors, or other individuals or companies to whom a mailing should be sent. Access includes a wizard to assist with creating a report that can be used to print names and addresses on a sheet of labels. The Label Wizard includes predefined formatting for all of the common label products available in office supply stores.

What You Will Do Sam Vestering, manager of North American Distributors, wants to send a mailing to all of the US distributors. You will generate a sheet of mailing labels to be used to address the envelopes.

Creating Mailing Labels

1 Open **4-WEDistributors** and enable the content, if necessary.

2 Click *US_Distributors* in the Tables group in the Navigation pane.

> Before you start the Label Wizard, select the table or query in the Navigation pane that contains the name and address fields.

3 Click the Create tab and then click the Labels button in the Reports group.

4 At the first Label Wizard dialog box, make sure *Avery* is selected next to *Filter by manufacturer*. Scroll up or down the list box below *What label size would you like?* and then click *5163* in the *Product number* column. *Note: Many labels are manufactured by Avery and you may need to scroll for a while to find the 5163 product number*.

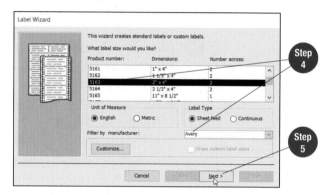

5 Click the Next button.

6 At the second Label Wizard dialog box, change the font size to *10* (if *10* is not currently selected) and then click the Next button.

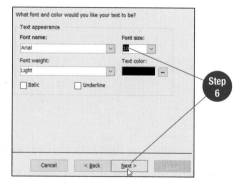

7 At the third Label Wizard dialog box, double-click *CompanyName* in the *Available fields* list box to move the field name to the first line in the *Prototype label* box.

> Access adds *CompanyName* in brace brackets on the first line of the prototype label. The prototype label represents a sample of the label that will be generated when you complete the wizard.

8 Press the Enter key to move the insertion point to the next line.

9 Double-click *StreetAdd1* in the *Available fields* list box and then press the Enter key.

10 Double-click *StreetAdd2* in the *Available fields* list box and then press the Enter key.

In Brief

Create Mailing Labels
1. Click table or query name in Navigation pane.
2. Click Create tab.
3. Click Labels button.
4. Select label product and click Next.
5. Select font characteristics and click Next.
6. Add fields including spacing and punctuation to prototype label and click Next.
7. Specify sort field and click Next.
8. Click Finish.

⑪ Double-click *City*, type a comma (,), press the spacebar, double-click *State*, press the spacebar, and then double-click *ZIPCode*.

⑫ Click the Next button.

⑬ At the fourth Label Wizard dialog box, double-click the *State* field name in the *Available fields* list box to move the field to the *Sort by* list box and then click the Next button.

> The labels will be sorted alphabetically by the *State* field.

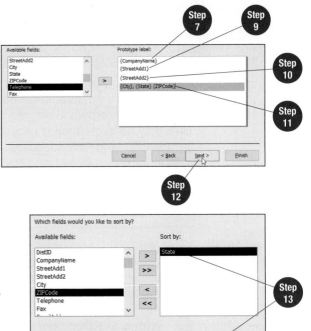

⑭ Click the Finish button at the last Label Wizard dialog box to accept the default report name *Labels US_Distributors*.

⑮ Click OK if a message box displays indicating some data may be lost because there is not enough horizontal space on the page.

> The labels report opens in Print Preview.

⑯ Scroll through the labels report and note that the labels are arranged in alphabetical order by the *State* field.

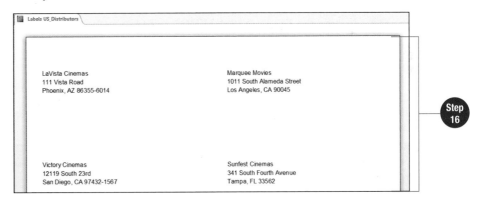

⑰ Print the labels report. Click OK if the message box reappears warning that some data may be lost.

⑱ Close the Labels US_Distributors report. ***Note: If the report displays in a Design view window after closing Print Preview, click the Close button at the top right of the work area.***

Check Your Work Compare your work to the model answer available in the online course.

Activity 4.11

Compacting and Repairing a Database; Backing Up a Database

After working with a database file for a period of time, the data can become fragmented because of records and objects that have been deleted. The disk space the database uses may be larger than is necessary. Compacting the database defragments the file and reduces the required disk space. Compacting and repairing a database also ensures optimal performance while using the file. The database can be set to compact automatically each time the file is closed. Databases should be regularly backed up to protect a business from data loss and to provide a historical record of tables before records are added, deleted, or modified. Access includes a backup utility to facilitate this process.

Worldwide Enterprises

What You Will Do You decide to compact the Worldwide Enterprises Distributors database file and then turn on the *Compact on Close* option so that the database is compacted automatically each time the file is closed. You will also create a backup copy of the database.

Tutorial

Compacting and Repairing a Database

Tutorial

Backing Up a Database

1. With **4-WEDistributors** open, click the Minimize button on the Title bar to reduce Access to a button on the taskbar.

   In the next steps, you will view the current file size of the database to determine the impact of compacting the database by comparing the before and after file sizes once you have compacted it.

2. Open File Explorer and then navigate to your AccessS4 folder.

3. If necessary, change to the Tiles view. To do this, click the View tab and then click the *Tiles* option in the Layout group.

   If the *Tiles* option is not visible, click the More button in the Layout group.

4. Locate the file named **4-WEDistributors** (the version that does not include a lock icon) and then write down the file size of the database.

   File size = _____

   Two files will appear in the file list for 4-WEDistributors. Notice one file displays with a lock icon. This file is used to lock records so that two users cannot request the same record at the same time.

5. Click the Access button on the taskbar.

6. Click the File tab and then click the Compact & Repair Database button at the Info backstage area.

   Compacting may take a few seconds.

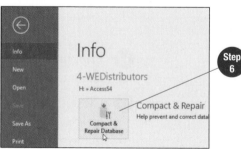

7. Click the File Explorer button on the taskbar.

8. Write down the new file size of **4-WEDistributors**. Notice that the amount of disk space used for the database is now lower.

   File size = _____

9. Close File Explorer.

In Brief

Compact and Repair Database
1. Open database.
2. Click File tab.
3. If necessary, click Info tab.
4. Click Compact and Repair Database button.

Turn on Compact on Close Option
1. Open database.
2. Click File tab.
3. Click *Options*.
4. Click *Current Database*.
5. Click *Compact on Close*.
6. Click OK.

Back Up Database
1. Open database.
2. Click File tab.
3. Click *Save As*.
4. Double-click *Back Up Database*.
5. Click Save.

10 With **4-WEDistributors** open, click the File tab and then click *Options*.

11 At the Access Options dialog box, click the *Current Database* option in the left panel.

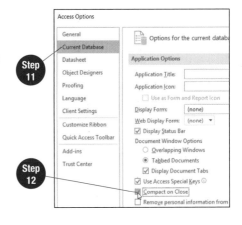

12 Click the *Compact on Close* check box in the *Application Options* section to insert a check mark.

13 Click OK to close the Access Options dialog box and then click OK at the message box that informs you that you must close and reopen the current database for the option to take effect.

14 Click the File tab and then click the *Save As* option.

15 At the Save As backstage area, double-click *Back Up Database* in the *Advanced* section.

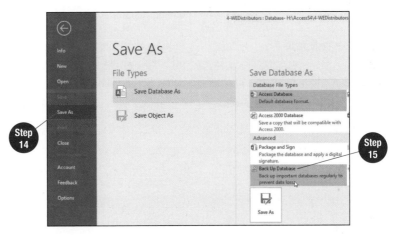

16 At the Save As dialog box, click the Save button to accept the default settings.

> By default, Access saves the backup copy of the database using the original file name with the current date appended to the end of the name. The backup copy is stored in the same location as the original database unless you change the drive and/or folder before saving.

17 Close **4-WEDistributors**.

In Addition

Editing Database Properties

Use the Database Properties dialog box (shown at the right) to store additional information about a database. For example, information can be added in the *Comments* text box about the data history in the backup copy of a database. To do this, open the backup copy of the database, click the File tab, and then click the <u>View and edit database properties</u> hyperlink in the right panel.

Features Summary

Feature	Ribbon Tab, Group	Button
aggregate functions	Query Tools Design, Show/Hide	Σ
back up database	File, Save As	
compact and repair database	File, Info	
conditional formatting in form or report	Form Layout Tools Format, Control Formatting OR Report Layout Tools Format, Control Formatting	
Crosstab Query Wizard	Create, Queries	
Find Duplicates Query Wizard	Create, Queries	
Find Unmatched Query Wizard	Create, Queries	
Group, Sort, and Total pane	Report Layout Tools Design, Grouping & Totals	
Label Wizard	Create, Reports	
Property Sheet task pane	Form Layout Tools Design, Tools OR Report Layout Tools Design, Tools	
sort a record in a form in ascending order	Home, Sort & Filter	

Word, Excel, and Access

Data Files

Before beginning section work, copy the IntegratingS2 folder to your storage medium and then make IntegratingS2 the active folder.

Skills

- Export Access data in a table to Excel
- Export Access data in a table to Word
- Export Access data in a report to Word
- Import Excel data to a new Access table

- Link data between an Excel worksheet and an Access table
- Edit linked data

Projects Overview

Export grades from an Access table to an Excel worksheet. Import grades from an Excel worksheet into an Access database table. Link grades between an Excel worksheet and an Access database table.

Export data on US Distributors from an Access table to a Word document. Export data on Canadian Distributors from an Access report to a Word document.

Export data on inventory from an Access table to a Word document.

Export data on costume inventory from an Access table to an Excel worksheet. Import data on costume design hours from an Excel worksheet into an Access table.

Link data on booking commissions between an Excel worksheet and an Access table and then update the data in the Excel worksheet.

The online course includes additional training and assessment resources.

One of the advantages of a suite like Microsoft Office is the ability to exchange data between programs. Access, like the other programs in the suite, offers a feature to export data into other programs—in this case, Excel and Word. Export data to Excel using the Excel button in the Export group on the External Data tab. An Access object such as a table, form, or query can be exported.

NIAGARA
PENINSULA
COLLEGE

What You Will Do   You are Katherine Lamont, Theatre Arts Division instructor at Niagara Peninsula College. You want to work on your grades for your AC-215 class over the weekend and you do not have Access installed on your personal laptop. You decide to export your Access grading table to Excel.

1 Open **2-NPCStudentGrades** from the IntegratingS2 folder and then enable the content.

2 Click the down arrow in the upper right corner of the Navigation pane and then click *Object Type* at the drop-down list.

3 Click the *StudentGradesAC215-03* query.

4 Click the External Data tab.

5 Click the Excel button ▦ in the Export group.

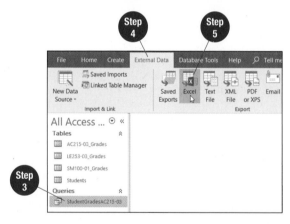

6 At the Export - Excel Spreadsheet dialog box, click the Browse button.

7 At the File Save dialog box, navigate to the IntegratingS2 folder on your storage medium.

8 Click in the *File name* text box, press the Home key to move the insertion point to the beginning of the file name, type 2- (the file name should appear as **2-StudentGradesAC215-03**), and then click the Save button.

> The name of the query was automatically inserted in the *File name* text box.

In Brief

Export Access Table, Form, or Query to Excel
1. Open database.
2. Click object in Navigation pane.
3. Click External Data tab.
4. Click Excel button in Export group.
5. At Export - Excel Spreadsheet dialog box, click Browse button.
6. At File Save dialog box, navigate to folder.
7. Click Save button.
8. Click options at Export - Excel Spreadsheet dialog box.
9. Click OK.

9 At the Export - Excel Spreadsheet dialog box, click the *Export data with formatting and layout* check box to insert a check mark.

10 Click the *Open the destination file after the export operation is complete* check box to insert a check mark and then click OK.

Excel opens with the grades from the query in cells in the workbook.

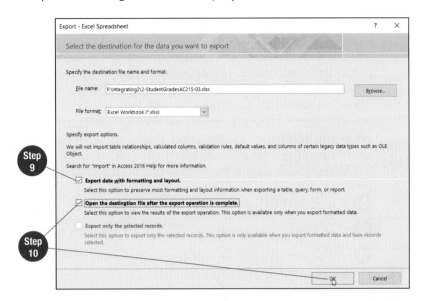

11 If Access is not active, click the Access button on the taskbar.

12 Click the Close button at the Export - Excel Spreadsheet dialog box.

13 Close **2-NPCStudentGrades**.

14 Click the Excel button on the taskbar.

15 In **2-StudentGradesAC215-03**, insert the following grades in the specified cells in the worksheet:

D4:	*C*
D8:	*B*
D10:	*F*
D14:	*A*
D16:	*C*
D18:	*B*

16 Save, print, and then close **2-StudentGradesAC215-03**.

	A	B	C	D
1	Student_No	Last_Name	First_Name	Grade
2	111-785-156	Bastow	Maren	B
3	118-487-578	Andre	Ian	A
4	137-845-746	Knowlton	Sherri	C
5	138-456-749	Yiu	Terry	A
6	146-984-137	Rhodes	Tari	A+
7	157-457-856	Dwyer	Barbara	C
8	184-457-156	Van Este	Doranda	B
9	197-486-745	Koning	Jeffrey	D
10	198-744-149	Lysenko	Earl	F
11	211-745-856	Uhrig	Andrew	A
12	217-458-687	Husson	Ahmad	A+
13	221-689-478	Bhullar	Ash	D
14	229-658-412	Mysior	Melanie	A
15	255-158-498	Gibson	Kevin	A+
16	274-658-986	Woollatt	Bentley	C
17	314-745-856	Morgan	Bruce	C
18	321-487-659	Loewen	Richard	B
19	325-841-469	Clements	Russell	A

Check Your Work Compare your work to the model answer available in the online course.

In Addition

Exporting Limitations

A table, form, or query can be exported to Excel but not a macro, module, or report. If a table contains subdatasheets or a form contains subforms, each subdatasheet or subform must be exported to view them in Excel.

Activity 2.2   Exporting an Access Table to Word

Data can be exported from Access to Word in a manner similar to exporting data to Excel. To export data to Word, open the database, select the object, click the External Data tab, click the More button in the Export group, and then click *Word* at the drop-down list. At the Export - RTF File dialog box, make changes and then click OK. Word automatically opens and the data displays in a Word document that is automatically saved with the same name as the database object. The difference is that the file extension *.rtf* is added to the name rather than the Word file extension, *.docx*. An RTF file is saved in "rich-text format," which preserves formatting such as fonts and styles.

What You Will Do   Sam Vestering, the manager of North American Distribution for Worldwide Enterprises, needs information on US Distributors for an upcoming meeting. He has asked you to export the information from an Access database to a Word document.

1. Open **2-WEDistributors** from the Integratings2 folder and then enable the content.

2. Click the *US_Distributors* table in the Tables group in the Navigation pane.

3. Click the External Data tab, click the More button [icon] in the Export group, and then click *Word* at the drop-down list.

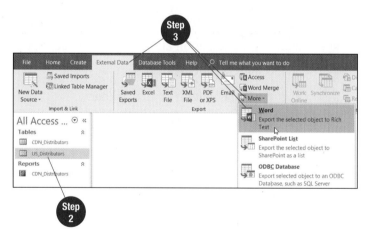

4. At the Export - RTF File dialog box, click the Browse button.

5. At the File Save dialog box, navigate to the IntegratingS2 folder on your storage medium.

6. Click in the *File name* text box, press Home to move the insertion point to the beginning of the file name, type 2- (the file name should appear as *2-US_Distributors*), and then click the Save button.

In Brief

Export Access Table to Word
1. Open database.
2. Click table in Navigation pane.
3. Click External Data tab.
4. Click More button in Export group.
5. Click *Word*.
6. At Export - RTF File dialog box, click Browse button.
7. At File Save dialog box, navigate to folder.
8. Click Save button.
9. Click options at Export - RTF File dialog box.
10. Click OK.

7 At the Export - RTF File dialog box, click the *Open the destination file after the export operation is complete* check box and then click OK.

Microsoft Word opens and the information for US Distributors displays in a document.

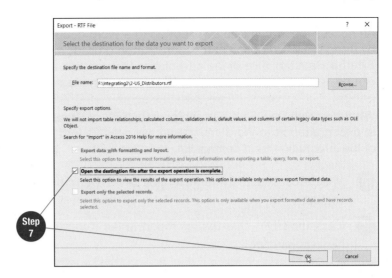

8 If Access is not the active program, click the Access button on the taskbar.

9 Click the Close button at the Export - RTF File dialog box.

10 Close **2-WEDistributors**.

11 If necessary, click the Word button on the taskbar.

12 Change to landscape orientation by clicking the Layout tab, clicking the Orientation button in the Page Setup group, and then clicking *Landscape* at the drop-down list.

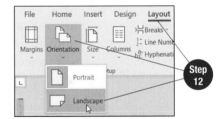

13 Click in any cell in the table.

14 Autofit the contents by clicking the Table Tools Layout tab, clicking the AutoFit button in the Cell Size group, and then clicking *AutoFit Contents* at the drop-down list.

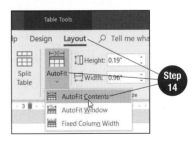

15 Save, print, and then close **2-US_Distributors**.

Check Your Work Compare your work to the model answer available in the online course.

In Addition

Adjusting a Table

In this activity, the Word table was adjusted to the cell contents. The table AutoFit feature contains three options for adjusting table contents as described at the right.

Option	Action
AutoFit Contents	adjusts table to accommodate table text
AutoFit Window	resizes table to fit within window or browser; if browser changes size, table size automatically adjusts to fit within window
Fixed Column Width	adjusts each column to fixed width using current widths of the columns

Activity 2.3 Exporting an Access Report to Word

An Access report can also be exported to a Word document. Export a report to Word by using the *Word* option at the More button drop-down list in the Export group on the External Data tab. One of the advantages to exporting a report to Word is that formatting can be applied to the report using Word formatting features.

What You Will Do Sam Vestering needs a list of Canadian Distributors. He has asked you to export a report to Word and then apply specific formatting to the report. He needs some of the information for a contact list.

1 Make Access active and then open **2-WEDistributors**. If necessary, enable the content.

2 Click the *CDN_Distributors* report in the Reports group in the Navigation pane.

3 Click the External Data tab, click the More button in the Export group, and then click *Word* at the drop-down list.

4 At the Export - RTF File dialog box, click the Browse button.

5 At the File Save dialog box, navigate to the IntegratingS2 folder on your storage medium.

6 Click in the *File name* text box, press Home to move the insertion point to the beginning of the file name, type 2- (the file name should appear as *2-CDN_Distributors*), and then click the Save button.

7 At the Export - RTF File dialog box, click the *Open the destination file after the export operation is complete* check box and then click OK.

   Microsoft Word opens and the Canadian Distributors report displays in a document.

8 If Access is not the active program, click the Access button on the taskbar.

9 Click the Close button at the Export - RTF File dialog box.

10 Close **2-WEDistributors**.

11 If necessary, click the Word button on the taskbar.

12 Convert the text to a table. To begin, click the Show/Hide ¶ button ¶ in the Paragraph group on the Home tab.

13 Move the insertion point to the left of the tab symbol that displays immediately left of the text *CompanyName*, press F8, and then press Ctrl + End.

   The keyboard shortcut F8 turns on the select mode.

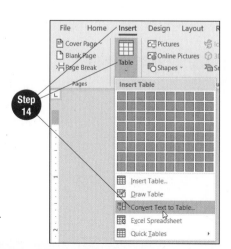

14 Click the Insert tab, click the Table button ▦, and then click *Convert Text to Table* at the drop-down list.

In Brief

Export Access Report to Word
1. Open database.
2. Click report in Navigation pane.
3. Click External Data tab.
4. Click More button in Export group.
5. Click *Word*.
6. At Export - RTF File dialog box, click Browse button.
7. At File Save dialog box, navigate to folder.
8. Click Save button.
9. Click options at Export - RTF File dialog box.
10. Click OK.

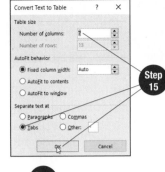

15 At the Convert Text to Table dialog box, make sure 7 displays in the *Number of columns* text box, make sure *Tabs* is selected in the *Separate text at* section, and then click OK.

16 Turn off the display of nonprinting characters by clicking the Show/Hide ¶ button in the Paragraph group on the Home tab.

17 Click in any cell in the first column (this column does not contain data) and then click the Table Tools Layout tab. Click the Delete button in the Rows & Columns group and then click *Delete Columns* at the drop-down list.

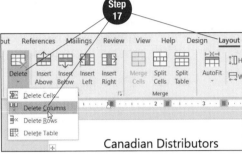

18 Change the left and right margins by clicking the Layout tab, clicking the Margins button in the Page Setup group, and then clicking *Normal* at the drop-down list.

19 Click the Table Tools Layout tab, click the AutoFit button in the Cell Size group, and then click *AutoFit Contents* at the drop-down list.

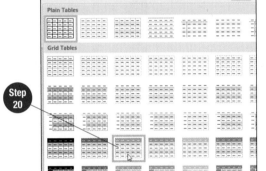

20 Click the Table Tools Design tab, click the More Table Styles button in the Table Styles group, and then click *Grid Table 4 - Accent 2* in the *Grid Tables* section. (The location on the grid may vary.)

21 Click in the title *Canadian Distributors* and then click the Center button in the Paragraph group on the Home tab.

22 Save, print, and then close **2-CDN_Distributors**. Close Word.

Check Your Work Compare your work to the model answer available in the online course.

In Addition

Merging Access Data with a Word Document

Word includes a Mail Merge feature for creating letters and envelopes, and much more, with personalized information. Generally, a merge requires two documents—the *data source file* and the *main document*. The data source contains the variable information that will be inserted in the main document. Create a data source file in Word or use data from an Access table. When merging Access data, either type the text in the main document or merge Access data with an existing Word document. To merge data in an Access table, open the database and then click the table in the Navigation pane. Click the External Data tab and then click the Word Merge button in the Export group. Follow the steps presented by the Mail Merge wizard.

Activity 2.4   Importing Data into an Access Table

In the previous three activities, Access data was exported to Excel and Word. Data from other programs also can be imported into an Access table. For example, data from an Excel worksheet can be imported that creates a new table in a database file. Data in the original program is not connected to the data imported into an Access table. If changes are made to the data in the original program, those changes are not reflected in the Access table.

NIAGARA PENINSULA COLLEGE

What You Will Do   You are Gina Simmons, theatre arts instructor. You have recorded grades in an Excel worksheet for students in the Beginning Theatre class. You want to import those grades into a database.

1   Make Access active and then open **2-NPCStudentGrades**. If necessary, enable the content.

2   Click the External Data tab.

3   Click the New Data Source button 🖳 in the Import & Link group, click *From File* at the drop-down list, and then click *Excel* at the side menu.

4   At the Get External Data - Excel Spreadsheet dialog box, click the Browse button.

5   At the File Open dialog box, navigate to your IntegratingS2 folder and then double-click *NPCBegThGrades*.

6   At the Get External Data - Excel Spreadsheet dialog box, click OK.

7   At the first Import Spreadsheet Wizard dialog box, insert a check mark in the *First Row Contains Column Headings* check box and then click the Next button.

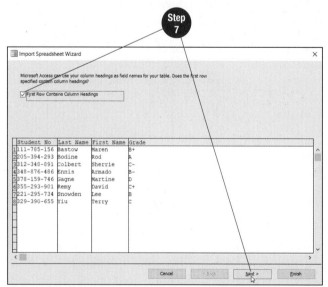

8   At the second Import Spreadsheet Wizard dialog box, click the Next button.

9 At the third Import Spreadsheet Wizard dialog box, click the *Choose my own primary key* option (this inserts *Student No* in the text box located to the right of the option) and then click the Next button.

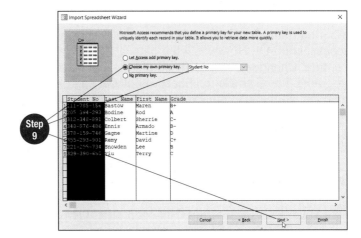

10 At the fourth Import Spreadsheet Wizard dialog box, type BegThGrades in the *Import to Table* text box and then click the Finish button.

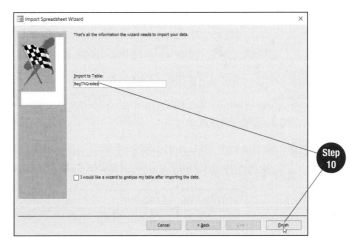

11 At the Get External Data - Excel Spreadsheet dialog box, click the Close button.

12 Open the new table by double-clicking *BegThGrades* in the Navigation pane.

13 Print and then close the BegThGrades table.

14 Close **2-NPCStudentGrades**.

Check Your Work Compare your work to the model answer available in the online course.

In Addition

Importing or Linking a Table

Data from another program can be imported into an Access table or the data can be linked. Choose the method depending on how the data will be used. Consider linking an Excel file instead of importing to keep data in an Excel worksheet but use Access to perform queries and create reports. In Access, linked data can only be updated in one direction. Once an Excel file is linked to Access, the data cannot be edited in the Access table. The data in the Excel file can be updated and the changes are reflected in Access, but the data cannot be updated within Access.

Activity 2.5   Linking Excel Data to an Access Table

Imported data is not connected to the source program. If the data will only be used in Access, import it. However, if the data will be updated in a program other than Access, link the data. Changes made to linked data are reflected in both the source and the destination programs. For example, an Excel worksheet can be linked with an Access table and, when changes are made in the Excel worksheet, the changes will be reflected in the Access table.

NIAGARA PENINSULA COLLEGE

What You Will Do   You are Cal Rubine, theatre arts instructor at Niagara Peninsula College. You record students' grades in an Excel worksheet and link the grades to an Access database table. With the data linked, changes you make to the Excel worksheet are reflected in the Access table.

1. Make Excel active and then open **NPCTRA220**.

2. Save the workbook with the name **2-NPCTRA220**.

3. Print and then close **2-NPCTRA220**.

4. Make Access active and then open **2-NPCStudentGrades**.

5. Click the External Data tab, click the New Data Source button in the Import & Link group, click *From File* at the drop-down list, and then click *Excel* at the side menu.

6. At the Get External Data - Excel Spreadsheet dialog box, click the Browse button.

7. Navigate to the IntegratingS2 folder on your storage medium and then double-click **2-NPCTRA220**.

8. At the Get External Data - Excel Spreadsheet dialog box, click the *Link to the data source by creating a linked table* option and then click OK.

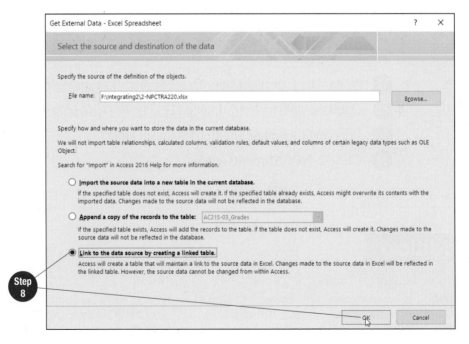

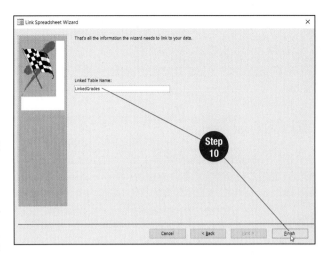

9 At the first Link Spreadsheet Wizard dialog box, make sure the *First Row Contains Column Headings* check box contains a check mark and then click the Next button.

10 At the second Link Spreadsheet Wizard dialog box, type LinkedGrades in the *Linked Table Name* text box and then click the Finish button.

11 At the message stating that the link is finished, click OK.

Access uses different icons to represent linked tables and tables that are stored in the current database. Notice the icon that displays before the LinkedGrades table.

12 Open the new LinkedGrades table in Datasheet view.

13 Close the LinkedGrades table.

14 Make Excel the active program and then open **2-NPCTRA220**.

15 Make cell E2 active, click the AutoSum button arrow in the Editing group, click *Average* at the drop-down list, and then press the Enter key.

This inserts *3.00* in cell E2.

16 Copy the formula in cell E2 down to the range E3:E9.

17 Save, print, and then close **2-NPCTRA220**.

18 Close Excel.

19 With **2-NPCStudentGrades** open, open the LinkedGrades table and notice that the worksheet contains the average amounts.

20 Print and then close the table.

21 Close **2-NPCStudentGrades** and then close Access.

	A	B	C	D	E
1	Student No	Student	Midterm	Final	Average
2	111-785-156	Bastow, M.	3.25	2.75	3.00
3	359-845-475	Collyer, S.	1.50	1.00	1.25
4	157-457-856	Dwyer, B.	3.50	3.50	3.50
5	348-876-486	Ennis, A.	2.25	2.00	2.13
6	378-159-746	Gagne, M.	3.00	3.50	3.25
7	197-486-745	Koning, J.	2.75	2.50	2.63
8	314-745-856	Morgan, B.	3.75	3.00	3.38
9	349-874-658	Retieffe, S.	4.00	3.50	3.75
10					

Step 16

 Check Your Work Compare your work to the model answer available in the online course.

In Addition

Deleting the Link to a Linked Table

To delete the link to a table, open the database and then click the table in the Navigation pane. Click the Home tab and then click the Delete button in the Records group. At the question asking if you want to remove the link to the table, click Yes. Access deletes the link and removes the table's name from the Navigation pane. When a linked table is deleted, the information Access uses to open the table is deleted, not the table itself. The same table can be linked again, if necessary.